FOOD MACHINERY
For the Production of Cereal Foods, Snack Foods and Confectionery

ELLIS HORWOOD SERIES IN FOOD SCIENCE AND TECHNOLOGY

Editor-in-Chief: I. D. MORTON, Professor and formerly Head of Department of Food and Nutritional Science, King's College, London
Series Editors: D. H. WATSON, Ministry of Agriculture, Fisheries and Food, and
M. J. LEWIS, Department of Food Science and Technology, University of Reading

Food Biochemistry   C. Alais & G. Linden
Fats for the Future   R.C. Cambie
Food Handbook   C.M.E. Catsberg & G.J.M. Kempen-van Dommelen
Determination of Veterinary Residues in Food   N.T. Crosby
Food Machinery: For the Production of Cereal Foods, Snack Foods and Confectionery   Ling-Min Cheng
Food Policy Trends in Europe: Nutrition, Technology, Analysis and Safety   H. Deelstra, M. Fondu,
   W. Ooghe & R. van Havere
Principles and Applications of Gas Chromatography in Food Analysis   M.H. Gordon
Nitrates and Nitrites in Food and Water   M.J. Hill
Technology of Biscuits, Crackers and Cookies, 2nd Edition   D.J.R. Manley
Modified Atmosphere Packaging of Food   B. Ooraikul & M. E. Stiles
Feta and Related Cheeses   R.K. Robinson & A.Y. Tamime
Vitamins and Minerals   M. Tolonen
Applied Human Nutrition: For Food Scientists and Home Economists   A.F. Walker

Forthcoming titles

Traditional Fermented Foods   M.Z. Ali & R.K. Robinson
Food Microbiology, Volumes 1 & 2   C.M. Bourgeois, J.F. Mescle & J. Zucca
Technology of Meat and Meat Products   J. P. Girard
Dairy Technology   A. Grandison, M.J. Lewis & R.A. Wilbey
Separation Processes: Principles and Applications   A. Grandison & M.J. Lewis
Microbiology of Chilled and Frozen Foods   W.F. Harrigan
Food Technology Data   M.J. Lewis
Education and Training in Food Science: A Changing Scene   I.D. Morton
Food: Production, Preservation and Safety   P. Patel
Natural Toxicants in Food   D.H. Watson
Food Container Corrosion   Wiese *et al.*
Chilled Foods: A Comprehensive Guide   Dennis & Stringer
Food Process Modelling: Relating Food Process Design, Food Safety and Quality   Niranjin & de Alwis

FOOD MACHINERY
For the Production of
Cereal Foods, Snack Foods
and Confectionery

LING-MIN CHENG
Associate Professor
The Faculty of Food Machinery
Hei Lonjiang Commercial College, China

ELLIS HORWOOD
NEW YORK   LONDON   TORONTO   SYDNEY   TOKYO   SINGAPORE

First published in 1992 by
ELLIS HORWOOD LIMITED
Market Cross House, Cooper Street,
Chichester, West Sussex, PO19 1EB, England

A division of
Simon & Schuster International Group
A Paramount Communications Company

Printed and bound in Great Britain
by Hartnolls, Bodmin, Cornwall

**Exclusive distribution by Van Nostrand Reinhold (International),
an imprint of Chapman & Hall, 2–6 Boundary Row, London SE1 8HN**

Chapman & Hall, 2–6 Boundary Row, London SE1 8HN, England

Van Nostrand Reinhold Inc., 115 5th Avenue, New York, NY 10003, USA

Nelson Canada, 1120 Birchmont Road, Scarborough, Ontario M1K 5G4, Canada

Chapman & Hall Japan, Thomson Publishing Japan, Hirakawacho Nemoto Building, 7F, 1-7-11 Hirakawa-cho, Chiyoda-ku, Tokyo 102, Japan

Chapman & Hall Australia, Thomas Nelson Australia, 102 Dodds Street, South Melbourne, Victoria 3205, Australia

Chapman & Hall India, R. Seshadri, 32 Second Main Road, CIT East, Madras 600 035, India

Rest of the world:
Thomson International Publishing, 10 Davis Drive, Belmont, California 94002, USA

British Library Cataloguing in Publication Data

Ling-Min Cheng
Food machinery: for the production of cereal foods, snack foods and confectionery. — (Ellis Horwood series in food science and technology)
I. Title II. Series
664.0028
ISBN 0–7476–0071–6

Library of Congress Cataloging-in-Publication Data

Cheng, Ling-Min, 1945–
Food machinery: for the production of cereal foods, snack foods, and confectionery / Ling-Min Cheng.
p. cm. — (Ellis Horwood series in food science and technology)
Includes bibliographical references and index.
ISBN 0–7476–0071–6
1. Food processing machinery. I. Title. II. Series.
TP373.C29   1992
664′.752′028–dc20
91–45189
CIP

Table of contents

Author's preface

Formerly, snacks, biscuits, bread, cakes, and confectionery were produced on a small scale, and a lot of labour was employed. Nowadays, food production has developed from a craft to a technological industry in both developed and developing countries. The mechanization of food production has made it possible to control the food quality at a unified and higher level, to reduce labour cost and decrease the risk of contamination, to avoid many laborious, dull, and repetitive jobs, and to obtain a higher processing efficiency. With the development of the large-scale food industry, more trained people with adequate understanding of food processing, not only technologically but also mechanically, are needed to keep up with the rising standards of the consumer market.

Many books on food technology have been published in recent years, but few on food machinery with special reference to snacks, biscuits, bread, cookies, cakes, and confectionery. This book is oriented to form a bridge between food technology and food mechanics, between industrialists and professional engineers, between machine designers and operators, between food machinery sellers and buyers, as well as between food producers and consumers.

This book provides a general technical and mechanical background for the basic processing machinery for making snacks, baked goods, and confectionery. The areas covered include the type, construction, and processing principles as well as transmission systems. Many illustrations and data tables, of up-to-date commercially available food machinery, give a clear objective impression to the readers either with or without the knowledge and experience of the snacks, baked goods, and confectionery industries. They should also help them to easily understand the functions of the machines, at a technical level that will enable correct and full use and a wise choice to be made when deciding which equipment to purchase.

The book commences with mixing and beating machines which are always situated at the beginning of the processing lines and are used as preparation equipment for food production. Chinese food occupies a bright and colourful place in the flower garden of the Worlds delicious food. It is a part of China's ancient civilization and long history. In recent years, it is attracting more and more attention and is becoming increasingly popular throughout the World. To taste Chinese food is also a pleasure for everyone, so that some typical Chinese snacks: steamed buns, dumplings, and Hun Tun, are introduced here with their technology, machines, and even manual methods in Part II. Important bread-making machines, dividers and rounders, are described in Part III, while Part IV is oriented toward biscuit production equipment, including sheeters, laminators, and reciprocating and rotary cutters, followed by the description of cookie-making machines, rotary moulders, and extruders in Part V. Cakes are often referred to as encrusted bakery food, and their processing equipment is represented by the Rheon encrusting machine described in Part VI. Since dough piece formation is normally followed by the baking process, baking ovens are discussed in the following Part VII which discusses heating principles, oven types, and the popular electric tunnel oven construction as well as its basic design. In Part VIII, important candy making machines, cookers, batch formers, rope sizers, candy makers, and candy mould depositing machines are described.

Sugar massecuites and some doughs are typical plastic materials, while most flour doughs are very complicated viscoelastic materials which often do not need a complex mechanism but rather an ingenious design, taking advantage of their rheological properties resulting in optimum food products. The complicated properties of food materials make food machinery design a rather complex subject requiring not only the knowledge necessary for common machinery design but also a knowledge of food engineering and technology, food science, food rheology, food hygiene, and machinery moulding skills. The final part, Part IX, emphasizes the basic characteristics of food machinery design for those who have a background of the designing similar machines but are specially interested in food machinery design.

Thanks are due to my supervisor, Professor I. D. Morton, Editor-in-chief of this book, for his encouragement and his help with my English, as well as his editing, thanks also to my superiors who offered me chances to visit and undertake further study in China and abroad on the food industry and food science, my colleagues who have helped me whenever needed, and particularly my husband for his unfailing support and encouragement.

The author is grateful to all the people, factories, companies, and organizations who have supplied information and illustrative materials and have given permission for their use in this book. Space limitations prevent the listing of them all. The following reviewed one or more chapters:

Mr Stephen Cook and Mr Mark Davies	Baked Products Division & Confectionery Division, APV Baker, UK.
Mr Peter Haussmann	The Bear Varimixer Company, Denmark.
Mr K. Kintscher	Robert Bosch GmbH, Germany.
Mr Rob Coppejans	Ter Braak BV, The Netherlands.
Mr Ar Dar Chi	Rheon Automatic Machinery Co. Ltd, Japan.
Mr Derek J. Benke	Rieckermann (HK) Ltd., Vicars Group Ltd, UK.
Mr. Bin-xiang Wang	Harbin Confectionery & Drink Corporation, China.

Mr Hua-jan Zhang and Mr Bin-sheng Wang	Harbin Commercial Machinery General Factory, China.
Mr Qing-ping Gu and Mr Ben Xu	Wu Xi Supply and Marketing Machinery Plant, China.
Mr Tong-gen Xu	Dong Tai Second Light Industry Machinery Factory, China.
Mr Zhen-fa Qiou	Lai Zhou City Food Machinery Factory, China.
Mr Chang-jiu Song	Harbin Food Machinery Factory, China.
Mr Quan-fu Dong	Shang Hai Feng Xian Cereal Machinery Factory, China.
Mr Chao-jiang Peng	Ru Gao Food Machinery Factory, China.
Mr Rui Ma and Mr Yi-bing Shi	Hei Long Jiang Commercial College, China.

Bepex GmbH Hutt, Germany.
Hai Yi Tao (Height) Inc., Japan.
OSHIKIRI Machinery Ltd, Japan.
Latini Machine Company, Inc., U.S.A.
Xi An Food (YIN SHI) Machinery Factory, China.
Shang Hai Food Machinery Factory, China.
Shang Hai Food (YIN SHI) Machines and Tools Factory, China
Jiang Yin Commercial Machinery Factory, China.
Xin Xiang County Food Machinery Factory, China.
Xin Xiang City Food Machinery Plant, China.
VEB KOMBINAT NAGEMA, Germany.

The author would also like to express her appreciation to the following publishers for granting permission to reproduce some tables and illustration in the book.
Academic Press, Inc., USA, New York
Van Nostrand Reinhold, USA, New York
Harper Collins, UK, London
Mir Publishers, USSR, Moscow

Ling-min Cheng
1991

Part I

Beating and mixing machines

1

Beating machines

1.1 INTRODUCTION

The term 'beating machines', or 'beaters' for short, derives from the use of a beating technique for making egg albumen emulsions. Beating machines are characterized by their unique function for making foams through planetary beating actions at high speed (about 100 to 300 rev/min is common), while some even rotate above 500 rev/min. To differentiate them from ordinary mixers, the term 'beating machines' is employed in this chapter, even though they can also mix doughs and other mixes.

Since their introduction, the range of applications of beating machines has increased, not only in the food field but in the chemical and pharmaceutical industries. For food production, they can be used in kitchens in restaurants, hotels, hospitals, bakeries, confectioneries, etc.

The early beating machines were generally used to make foams which require the generation of mechanical energy within the liquid by means of beating or whipping the liquid or solid to the continuous phase, with gas in small bubbles as the dispersed phase, such as in candy foam and sponge batter preparation. Simple and small beating tools with a vigorous mode of revolution were used. With time, beaters have been developed in various sizes, with many types of beating tools, to perform not only beating (emulsifying) but also kneading and stretching actions. Thus, beating machines, nowadays, are not only used for candy foams and sponge batters but also for mixing high-viscosity mixtures such as confectionery massecuites and bread doughs for either the Chorleywood or common fermentation process.

This development is also seen in advanced control systems, some of which are equipped with computers to allow 'finger touch' operation, and in the introduction of beating under pressure rather than at natural atmospheric pressure.

According to the orientation of the beating shaft, beaters are classified as horizontal and vertical beaters. Vertical beaters are widely used because of their powerful and stable output, simple but durable mechanism, and especially for the wide range of their application to various raw materials such as powders, liquids, pastes and high-viscosity mixtures. Horizontal beaters are seldom seen nowadays.

1.2 CONSTRUCTION

1.2.1 General

A beating machine comprises a mixing bowl, a planetary head driven by a motor via a transmission system, and the frame in which the internal mechanisms are enclosed. This frame has safety covers which provide easy access for cleaning and maintenance. To ensure hygienic conditions, all surfaces in contact with food materials are made of noncorrosive food-compatible materials.

Vertical beating machines are generally designed in two different configurations. One type, as shown in Fig. 1.1, is constructed as a symmetrical structure with the planetary head mounted under the bridge and supported by the symmetrical frames. Another type is shown in Fig. 1.2, in which the planetary head is fixed under the cantilevered frame.

For working stability, the former type (Fig. 1.1) is more suitable for heavy duty operation with a bowl size up to 600 litres and the latter type (Fig. 1.2) is more suitable for lighter duty operation with a bowl size under 200 litres. For heavy duty work, the machines are often equipped with two mixing tools, while for light duty one mixing tool is usually enough. Tables 1.1 and 1.2 give general specifications of the two types of machines.

The air consumption data in Table 1.1 indicate that the beating operation is under pressure (between 2 and 8 atmospheres) introduced via the air tube shown in Fig. 1.1. The bowl is raised from the position shown, and the fixed cover ensures totally enclosed and sealed mixing.

Pressure beating is a relatively new technique for making foams. Air is continuously absorbed by the mixture, with an air consumption up to 1000 litres per hour (Table 1.1). When the pressure is released, the fine air bubbles incorporated greatly expand. Pressure beating accelerates aeration, results in a five-fold reduction in beating time (7), and saves on ingredients.

Mixing can also be carried out under vacuum, by extracting air from the bowl, for special purposes. However, most beating operations perform quite well under normal

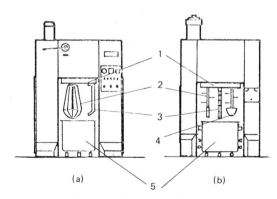

Fig. 1.1 Beating machines (a) with one mixing tool, (b) with two mixing tools (Courtesy of APV Baker, UK) (1) Planetary head, (2) mixing tool, (3) high-pressure air tube, (4) mixing tool, (5) mixing bowl, before lifting.

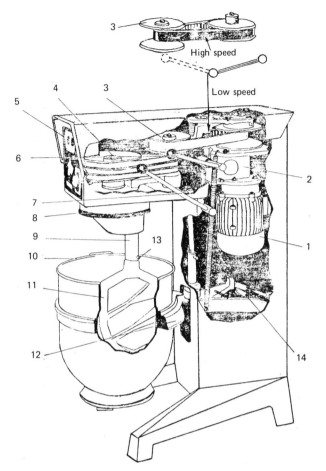

Fig. 1.2   Varispeed mixing machine (Courtesy of A/S Wodschow Co., Denmark) (1) Motor, (2) Worm-gear drive, (3) Varispeed drive, (4) Pulley, (5) Facia panel, (6) Gear lever, (7) Bowl-raising lever, (8) Planetary head, (9) Mixing shaft, (1) Mixing bowl, (11) Mixing (beating) tool, (12) Bowl arm, (13) Bayonet lock, (14), Automatic bowl-clamping mechanism.

atmospheric pressure conditions; they are referred to as natural beating or natural mixing.

1.2.2   Mixing tools

In making different foams, doughs, and other mixes, mixing tools perform several necessary actions, such as beating, blending, kneading, rolling, stretching, shearing, and even cutting.

A large selection of mixing tools are available for various types of food materials. They have a wide range of physical properties. Fig. 1.3 shows some commonly used tools. The different configurations are often known as whisks (whips), hooks, spades or flat beaters, and mixing arms.

Table 1.1   Specifications of some beating machines with symmetrical stand frame structure (Courtesy of APV Baker, UK.)

Bowl size (litre)	Approximate Dimensions (mm)			Weight (kg)	Motors (kW)			Air consumption litres/hour
	Height	Length	Width		Drive	Hoist	Compressor	at 1 atm
80	2030	1340	665	860	2.2	0.55	0.55	120
120	2215	1420	825	920	3	0.55	0.55	200
120 (+collar)	2265	1420	825	920	3	0.55	0.55	200
120 (DC)	2410	1420	825	950	2.2–6.7	0.55	—	200
120 Panel	1500	800	500	200	—	—	—	—
300	2960	1800	1150	2200	4.5–13.4	3	—	600
300 Special	3150	1800	1150	2190	9.3	3	—	600
300 Panel	1800	800	500	200	—	—	—	—
400	3500	1890	1580	3180	8.6–26	4.1	—	800
600	3950	2370	2020	4250	15–45	4.1	—	1000
400/600 Panel	1800	800	500	220	—	—	—	—

N.B. Min air pressure 2 atm.
     Max air pressure 8 atm.

Table 1.2   Specifications of some beating machines with cantilevered column structure (Courtesy of A/S Wodschow & Co., Denmark)

Bowl size (Litre)	20	30	40	60	100	150
Mixing tool min.	83	57	53	53	47	31
speed (rev/min) max.	373	311	294	288	257	174
Dimensions (mm)						
Length	634	820	840	980	1090	1300
Width	390	460	500	700	730	780
Height	740	1200	1200	1365	1415	1750
Motor (kW)	0.74	0.93	1.10	1.65	3.0	4.4
Net weight (kg)	96	156	163	233	303	650

Whisks are assemblies of a special tough stainless steel wire, either pear shaped (Fig. 1.3(1) and (4)), or wide at the top and coming to a point at the bottom, or cylindrical (Fig. 1.3(2)) or with cruciform cross-section (Fig. 1.3(3)). The wire assembly is fixed to the mixing shaft which is in turn mounted on the planetary driving shaft by means of bayonet lock to receive the revolving and rotating torques. Beating whisks can provide efficient whipping even of quantities as small as 20% of the maximum capacity of the bowl. The design of the whisk configuration is intended to give the maximum air incorporation and bubble-dividing actions. APV Baker claim that the pear shaped

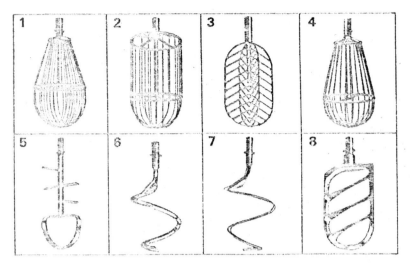

Fig. 1.3   Mixing tools (Courtesy of APV Baker, UK)

whisk (1) is suitable for whipping creams, meringues, light sponge batter, etc. The cylindrical whisk (2), made of heavy wire, is suitable for mixing creams, batters, custards, etc. The cruciform type (3) is suitable for butter cream, wafer creams, etc. and the pear-shaped whisk (4), also made of heavy wire, is for batters, heavier creams, light cake mixes, etc.

Blade and float spade mixing tools (Fig. 1.3(5) and (8)) are made of stainless steel and are shaped like propellers, giving remarkable vertical mixing effects which ensure a more uniform product such as heavier cake mixes, and short pastry in less time. Sometimes a plastic or rubber scraper is fitted into the edge of the tool to wipe the viscous products from the inside of the bowl.

The mixing arms, straight hook or spiral shaped, are usually made of stainless steel and are commonly used in dough mixing. The spiral mixing arm (Fig. 1.3(6)), tends to lift ingredients such as rotary biscuit dough and shortbread. The spiral arm (Fig. 1.3(7)) tends to press and stretch glutenous doughs for buns, rolls, etc.

1.2.2   Mixing bowl

The stainless steel mixing bowl is commonly designed with a cylindrical body and a hemispherical or flat bottom. The two parts can be welded together or made in one piece by a deep drawing die press. Bowl sizes are from less than 10 to more than 600 litres. For a small unit, the bowl is supported by bowl arms, and is raised by turning a handwheel and is clamped by a clamp lever. This brings pressure onto the wedge plane of the L-type strut band of the bowl, and in a single action, the bowl is automatically in its working position. For a large unit, the bowl is controlled automatically.

In some early types of beating machine, the mixing shaft can be raised by means of a handwheel–screw mechanism which allows the bowl to be rolled in under it. After the bowl is locked, the shaft lowered, beating can start.

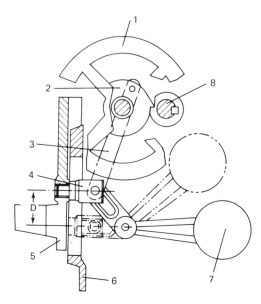

Fig. 1.4   Bowl-lift mechanism [15] (1) Handwheel, (2) Cam, (3) Connecting rod, (4) Sliding block, (5) Bowl arm, (6) Machine stand, (7) Balancer, (8) Pin.

A typical bowl-lift mechanism is shown in Fig. 1.4. By turning handwheel (1), the coaxial cam (2) actuates the sliding block (4) through the connecting lever (3) so that the bowl arm (5) is made to travel upward or downward vertically along the swallowtailed guideway on the machine stand (6). The raising distance D of the bowl is dependent on the eccentricity of the cam, which is commonly about 60–70 mm. When the raised part of cam (2) is engaged with the location pin (8), the bowl is lifted and locked to its upper limit, while balancer (7) produces a pushing force upward through the slide block pin to reduce the gravity on the bowl arm during raising and lowering.

Some bowls are lifted or lowered by means of a screw-nut mechanism. By turning the handwheel, the screw is made to revolve, so that the nut on the bowl arm moves up and down, producing the up-down motion of the bowl.

As shown in Figure 1.5, a stainless steel anti-splash ring is placed on the mixing bowl to prevent liquid and powder ingredients from splashing during beating.

1.2.3   Planetary head

Just as the Earth revolves around the Sun, so the mixing tools of the beating machine move in a planetary mode. This machine, adopting such a characteristic local motion, provides a high mixing performance without having any point during the operation.

The planetary head (Fig. 1.6) consists of a ring gear, rotary arm, beating (mixing) shaft, and planetary gear. The ring gear is fixed on the frame of the machine. The rotary arm receives its rotary motion from the drive system and delivers it to the planetary gear which engages with the ring gear at the same time. Thus the beating

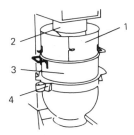

Fig. 1.5   Application of anti-splash ring (Courtesy of A/S Wodschow Co., Denmark) (1)
Anti-splash ring, (2) Planetary head, (3) Mixing bowl, (4) Bowl arm.

shaft not only receives its rotary motion around the vertical axis of the rotary arm and
the bowl, forming a revolution, but also rotates around its own axis at the same time.
Subsequently, the revolving and rotating of the beating (mixing) tool carried by the
shaft take place and gives the desired mixing, beating and whipping, or kneading
effect. For Fig. 1.6, the relationship between rotation and revolution of the mixing tool
can be expressed in the equation

$$n_{rot} = \left(1 - \frac{Z_{ring}}{Z_{planet.}}\right) n_{rev.} \tag{1.1}$$

where $n_{rot.}$ is the speed of the mixing tool rotation, $n_{rev.}$ is the speed of the tool
revolution or the planetary head speed, usually the mixing speed in machine
specifications, Z_{ring} is the number of teeth on the ring gear, and $Z_{planet.}$ is the number
of teeth on the planetary gear.

Equation (1.1) indicates that the difference in speed between rotation and revolution of the mixing tool depends on the difference in teeth number between the ring
gear and the planetary gear. Since the ring gear always has more than the planetary
gear, the speed of rotation is invariably greater than that of the revolution. That is, the
local mixing speed is always faster than the whole. The negative sign in the calculation
result means that the rotation and revolution of the mixing tool are in opposite
directions.

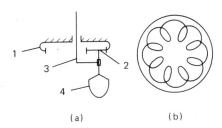

(a)   (b)

Fig. 1.6   Schematic of planetary head (a) and mixing tool orbit (b) [15] (1) Ring gear, (2)
Planetary gear, (3) Rotary arm, (4) Mixing tool.

1.3   TRANSMISSION SYSTEM

The transmission system of a beating machine comprises the planetary head, gear shift, and an electric motor. In most machines, the motor is built-in and installed at the lower part of the main body so that the centre of gravity is more stable, and less vibration occurs. This also provides a better appearance and better operation.

There are two ways of providing speed variation of the beaters. Some machines are designed with a gear transmission as shown in Fig. 1.7 to provide a choice of 3 to 5 speeds (Table 1.3), while others have a variable drive permitting selection of any desired speed within the limit of the equipment, as shown in Fig. 1.2. By turning the gear lever, the mixing (beating) speed is changed gradually from a very low minimum speed to a very high maximum speed (Table 1.2).

The gear transmission process is commonly accomplished by means of an electric motor, belting, and reduction and steering gears (Fig. 1.7). The motor delivers its torque and motion via a one-stage belting, to the triple gear assembly, reducing the speed and providing a choice of three speeds. Then the rotation is transferred by a set of bevel gears, changing the direction of rotation, to the rotary arm to make the mixing tool rotate and revolve in a planetary orbit. The transmission scheme is Scheme 1.1.

$$\text{Motor 19} \to \frac{\text{Pulley 1}}{\text{Pulley 2}} \to \text{Shaft I} \begin{cases} \dfrac{\text{Gear 8}}{\text{Gear 3}} \\[2ex] \dfrac{\text{Gear 9}}{\text{Gear 4}} \to \text{Shaft II} \\[2ex] \dfrac{\text{Gear 10}}{\text{Gear 5}} \end{cases}$$

$$\text{Shaft II} \to \frac{\text{Gear 7}}{\text{Gear 6}} \to \text{Shaft III} \to \frac{\text{Bevel gear 11}}{\text{Bevel gear 12}} \to \text{Shaft IV}$$

Shaft IV → Planetary head → Mixing tool rotates and revolves

Scheme 1.1.   Transmission scheme for a beating machine.

The varispeed drive consists of two movable expanding and contracting pulleys (stepless shifting mechanism), strong special V-belts which ensure efficient power transfer to the mixing tool and which can absorb all shocks and loads during the mixing operation, and an electric motor with the power values given in Table 1.2. The application of a stepless belt-pulley reducer makes the transmission route shorter and simpler. Its transmission scheme is shown in Scheme 1.2.

Motor with worm-gear box → Variable speed drive → Belt-pulley with fixed ratio → Planetary head → Mixing tool

Scheme 1.2.   Transmission scheme for a stepless beating machine.

Table 1.3   Typical speeds changed by geared transmission system

Model	A	B	C
Bowl size (litre)	20	40	60
Mixing tool (rotation) (rev/min)	106 147 213 325	69 109 154 242	49 86 98 174
Planetary head (revolution) (rev/min)	28 39 57 87	33 52 73 115	23 41 47 83

This drive system is also characterized by the wormgear box which provides an optional 'attachment drive' as well as the drive shaft to the varispeed drive for the planetary head. The attachment drive is located at the left-hand side of the machine. It permits connection of accessory equipment, such as a meat mincer or cutting/slicing/shredding/grating equipment.

The use of various types of accessory equipment make the beating machine a universal working station in kitchens, confectioneries, and bakeries.

The planetary head and the gearcase for the attachment of this machine are greased initially for several years of operation, with the use of a recently developed grease. In this way, the sealing structure of the planetary head is simplified and there is no leakage of the lubrication oil.

For the machines shown in Fig. 1.1, the transmission process can also be accomplished by systems similar to that described above. There will be two planetary gears to engage with the stationary ring gear for driving two mixing tools respectively. As they are fixed 180° out of phase, there will be no interference with one another during mixing.

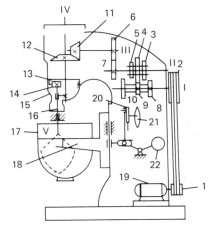

Fig. 1.7   Transmission system [15] (1,2) Pulley, (3–10) Spur gearing, (11,12) Bevel gears, (13) Ring gear, (14) Planetary gear, (15) Planetary head, (16) Mixing shaft, (17) Mixing bowl, (18) Bowl arm, (19) Motor, (20) Cam, (21) Handwheel, (22) Balancer. I, II, III, IV, V: Shaft numbers, as in Section 1.3, Transmission scheme.

1.4 PRODUCTION CAPACITY

The capacity of the beating machines is dependent upon the size of the bowl and the total cycle time including feeding, mixing and emptying.

Modern ingredient feed systems and discharge arrangements allow the auxiliary work to be done very quickly, and the high-efficient planetary mixing actions make it possible to achieve for different doughs, the short production cycles given in Table 1.4.

1.5 CONTROL AND SAFETY

There are three types of control system for the beating machines: manual control, servo-control, and finger touch control.

Manual control is used for most light-duty beating machines. It consists of START, STOP and EMERGENCY push-buttons as well as a timer located on the machine facia panel for setting beating time. The bowl clamping and speed adjustment can be achieved by the respective handles or handwheels generally located on the right-hand side of the machine for the convenience of right-handed operators (Fig. 1.2).

In the servo-control system, the handwheels or handles for speed adjustment and bowl lifting are replaced by SPEED UP AND DOWN, BOWL UP AND DOWN push-buttons which are located on the facia panel beside the timer and the buttons provided for manual control. The servo-control system can automatically reset the machine to low speed at the end of a working cycle to ensure a smooth start-up and to prevent accidental splash when starting at high speed. (A/S Wodschow & Co.).

The so-called 'finger touch control' is a programmable system. The computer memory allows for up to 25 programs with up to 9 speed/time steps per program. The computer displays all the machine functions on the facia panel in bright easy-to-read digital read-outs, giving program number, step number, speed, time and diagnostic data. If necessary, the machine can be provided with an optional computer interface for transmission of operational information to a PC or main frame computer for production control purposes, statistical calculations, transmission to headquarters, etc. (A/S Wodschow & Co. Denmark).

Table 1.4 Typical production step (APV Baker)

Bowl size (approximately)	200 litres	
	(kg)	(minutes)
Bread (Chorleywood)	105	4
Scotch puff pastry	105	$1\frac{1}{2}$
Sweet paste	110	3
Cold water paste	100	$2\frac{1}{2}$
Madeira cake	100	$4\frac{1}{3}$
Fruit cake	130	5
Cup cake	100	5

Working safety is extremely important for all beating machines since they are generally operated at high beating speed. Thus every type of control system ensures that the mixing bowl cannot be lowered until the beating (mixing) tool is stationary, and conversely, the machine cannot be started unless the bowl has been placed in position by means of limiter switches.

2

Mixers

2.1 INTRODUCTION

Mixing, in the snack and bakery field is an operation by which two or more materials such as flour, water, yeast, sugar, syrup, salt, minced meat, fruit, vegetables, or any other suitable foodstuff, are brought together to form a uniform mass through the application of external force imparted by mixing tools which are commonly referred to as agitators, mixing tools, blades, or beaters.

The machines performing the mixing functions are called mixers for short. They have been studied by many investigators and developed for many years. They are made in a wide range of styles, sizes, speeds, capacities, and construction materials for numerous food productions.

There are different ways to classify the mixers.

(1) According to the number of mixing spindles. There are single-spindle mixers and double-spindle or even triple-spindle mixers.
(2) According to their mixing speed. There are slow-speed mixers (less than 30 rev/min), high speed mixers (above 35 rev/min), and variable speed mixers.
(3) According to their operation mode. They can be classified into batch mixers and continuous mixers.
(4) According to the axis position of the mixing spindle from which the mixing arms receive torque and motion. They can be classified into vertical mixers and horizontal mixers. In this chapter, the machines will be discussed in terms of this classification.

Investigations show that horizontal mixers are still the dominant mixing equipment in today's modern bakery and snack industry, for they are of simple construction, simple in operation, and cheaper to run. They also have varied capacities and can be used for a wide variety of mixtures from a thin batter for cookie depositing to extremely tough dough for Chinese snack casing.

In the second place are the vertical mixers.

2.2 HORIZONTAL MIXERS

Horizontal mixers are characterized by having a horizontally located mixing spindle on which the mixing arms are fixed into the mixing bowl. Fig. 2.1 is a typical front view of this kind of mixer.

2.2.1 Construction

A typical horizontal mixer consists of a mixing bowl, one or two mixing spindles by which the mixing arm(s) is or are driven through transmission mechanisms, and a main frame made of either cast iron or unitary construction of heavy steel plate. One or two motors are mounted below for mixing and bowl tilting functions together with a facia control and an electric interlock system to prevent access when the machine is running.

There are two types of weighing systems: one is separate from the mixer; the other calculates the weight change of the complete mixer before and after the addition of an ingredient, the mixer being located on a suitable weighing scale or platform. In this case the mixer is often referred to as a weigh-mixer.

2.2.1.1 *Mixing bowl*

The bowl of the horizontal mixers is of trough-like design with a curved bottom (U-shaped in cross section) and flat ends. The bowl surfaces in contact with the dough are commonly of stainless steel or stainless clad steel. This is the usual construction for the bowl ends, where the bearings are fixed to support the mixing spindles. The bowls of large modern mixers are generally double-'skinned' in the form of a jacket through which chilled water or refrigerant can be circulated (Figure 2.2) to prevent the dough warming up to too high a temperature as a result of mixing friction.

To avoid flour and other ingredients splashing, especially at the beginning of mixing, and for safety as well as food hygiene, the bowl is always equipped with a lid

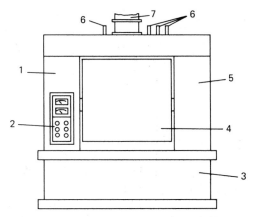

Fig. 2.1   Front view of a typical horizontal mixer (1) Gear box, (2) Facia control, (3) Frame, (4) Mixing bowl, (5) Gearbox, (6) Ingredients inlet, (7) Flour inlet.

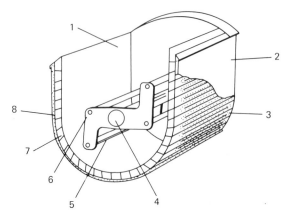

Fig. 2.2   Mixing bowl with jacket (Courtesy of Hai Yi Tao Inc., Japan) (1) Inside wall of bowl, (2) Outside wall of bowl, (3) Coolant, (4) Centre shaft, (5) Arm (roll), (6) Connection plate, (7) Jacket, (8) Heat insulation layer.

which is either removable or hinged for dough discharge and cleaning. For large mixers, the lid usually has provision for assisted ingredients feed.

There are two methods of dough discharging: by tilting the bowl (110° to 180°), or by mechanically sliding down the door in front of the stationary bowl to allow the dough to fall into an underlying hopper. For a ground-floor installation, the dough is often discharged into a dough tub which is usually fabricated in heavy gauge stainless steel and is supplied separately by the manufacturer.

The bowl-tilting operation is generally carried out by a worm-gear mechanism in which the worm-gear is fixed on the bowl sidewall (Figs 2.7 and 2.8).

Feeding of the bowl is carried out either manually for small mixers, or automatically through the corresponding pipes above the mixer and by means of a weighing system for large horizontal mixers.

Bowls are manufactured in a wide range of volumes which allow from a few kilograms up to 1500 kg of food materials to be mixed in them. The larger the bowl size, the greater the required power of the mixing motor, so that bigger batches of dough can be mixed, resulting in a higher rated capacity for the mixer. For most large mixers, the bowl is tilted by a separate reversible motor ranging from 0.75 to 2.26 kW.

2.2.1.2   Mixing arms

Mixing speed
The mixing operation is directly performed by the mixing arms, while its power is transmitted by its driving spindle (shaft or axle). That is, the speed of the mixing arm is dependent on the speed of its spindle. Horizontal mixers are designed in either a single or dual mixing speed mode. For the dual mode, its lower speed is half the rated maximum speed. As dough mixing is often carried out in two stages—blending of the ingredients, and developing the gluten—it is essential that the first stage should be accomplished at a lower speed (for example 36 rev/min) and the second stage at the .

rated speed (which will be 72 rev/min). Generally speaking, the machine with a mixing speed below 30 rev/min is referred to as a slow-speed mixer, and that with a speed above 35 rev/min as a high-speed mixer (see Table 2.1 and Table 2.2). Modern mixers commonly cover a wide range of speed variation from 20 to 145 rev/min or even up to more than 200 rev/min, which high speed allows a quick development of gluten elastic doughs by means of suitable mixing arms.

The slow-speed mixers are generally used in short and soft dough mixing since a much longer time would be needed for hard and bread doughs.

Form of the mixing arms
The mixing arms are designed in various configurations and cross-sections for different mixing functions such as blending, dispersing, beating, shearing, scraping,

Table 2.1 Specifications of some small-capacity horizontal mixers (Courtesy of Xin Xiang Food machinery Factory, China)

Model No.	1	2	3
Mixing speed (rev/min)	35	25	38
Batch capacity (kg)	75	27.5	25
Main motor (kW)	4	3	2.2
Bowl tilt motor (kW)	1.0		
Net weight (kg)	800	300	220
Dimensions (mm)	1180 × 670 × 1320	1180 × 630 × 960	860 × 500 × 910

Table 2.2 Specifications of some large 'shaftless' horizontal mixers (Courtesy of APV Baker, UK)

Model	1	2	3
Batch capacity (kg)	550 (soft dough) to 470 (hard dough)	850 (soft dough) to 725 (hard dough)	1100 (soft dough) to 1000 (hard dough)
Mixing speed (rev/min)	60/30	56/28	54/27
Dough discharge height (mm)	925	925	954
Dough discharge width (mm)	1050	1220	1366
Main motor	Standard 56/27 kW 1P22	Standard 75/37 kW 1P22	Standard 112/56 kW IP22
Bowl tilt motor	1.5 kW reversible	1.5 kW reversible	2.25 kW reversible
Dimension (mm)	3260 × 2260 × 2500	3580 × 2360 × 2700	3800 × 2600 × 2900
Net weight (kg)	6000	9000	11000
Jacket water	114 litres per minute at a pressure of 1.7 bar		

stretching, or kneading to form either a uniform mass or a dispersion or a solution, or aeration (that is, either a soft dough or hard dough, a sponge dough or batter or topping with other food material). Their locations in horizontal mixers are shown in Fig. 2.3.

Some mixing tools have a floral-hoop type, oval-type, or twisting-plate type (Fig. 2.4, a,b,c) and comprised only one or two loop-like arms without a centre shaft; they are referred to as 'shaftless' agitators or mixing arms. The corresponding machines are referred to as 'shaftless' mixers. Table 2.2. shows the specifications of some large 'shaftless' horizontal mixers.

'In this group, there are some other types of arm such as Z-type and S-type. Their cross-section is large to ensure strength. Their relatively complex configurations are commonly cast in one piece or are welded after forging. Attention should be paid to the coaxiality of the two sides of the arm during manufacturing to avoid severe trouble in the later mixing operation.

This type of mixer can be used for a wide range of doughs with different consistencies, from thin batter to extremely tough doughs, as the 'shaftless' arms are designed to achieve rapid dispersion of ingredients without any dead spots. But it is especially efficient in dealing with extensible doughs, since in their rotation orbit there is always a limited clearance from the bowl inner walls, which is beneficial in showing the dough to be stretched and kneaded repeatedly to form an oriented gluten network.

Some other mixing tools (agitators) comprise simple shaped arms and a centre shaft (Fig. 2.4, d and e). This kind of segmented construction is easy to manufacture and assemble, and its maintenance is lower than that of those described earlier. However, to deal with sticky doughs, this group of agitators are at a disadvantage since the tendency of sticky doughs is to adhere to the shaft, and the circular velocity at the centre shaft area is very low, resulting in a dead space and therefore improper mixing. Sometimes the centre shaft is covered by dough, layer upon layer. But if the centre shaft and arm are located as shown in Fig. 2.3, the problem can be reasonably well solved.

The pedal like arms (Fig. 2.4, e) can be either cast or forged. This kind of agitator is suitable for mixing short and soft doughs, since its configuration provides not only

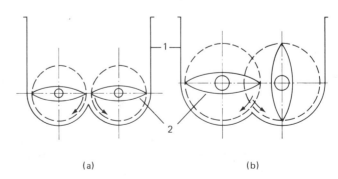

(a)   (b)

Fig. 2.3   Relative locations of mixing arms in horizontal mixers. (1) Mixing bowl, (2) Mixing arm.

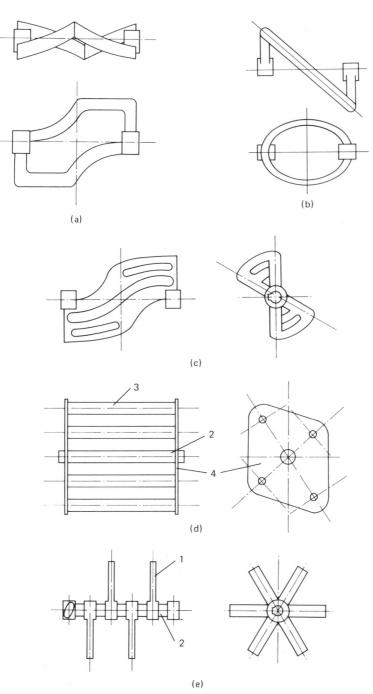

Fig. 2.4   Some types of agitator [45] (a) Floral-hoop-type, (b) Oval-type, (c) Twisting plate-type, (d) Cage-type with roll arm, (e) Pedal-type (1) Pedal arm, (2) Centre shaft, (3) Roll arm, (4) Connection plate.

blending but also shearing and cutting actions rather than stretching and kneading during mixing.

The roll arms ((3) in Fig. 2.4d) of the cage-type agitator revolve in their bearings during mixing, thus lessening the imparted force on the dough. This type of agitator, with four or six freely rotating rolls, can be used in a variety of dough mixings.

The location of the mixing arms
The horizontal mixers may be equipped with one or two driving spindles on which the agitators are mounted just clearing the lower part of the bowl during rotation. The gap between the inner surface of the bowl and the agitator edges is very important for dough mixing quality and power consumption. For two-spindle mixers, the bowl bottom would be more W-shaped to match the rotation orbit of the two agitators which can be located in parallel or in different phases to prevent interference with each other and consequent damage.

In Figure 2.3a the two agitators are in parallel, and rotate face to face. The clearance between their outer orbit of rotation is limited to prevent any dead spot in mixing. In this arrangement, there is no interference at all, no matter whether their speed is the same or not, as they are independent of each other. It is like a combination of two single-spindle mixers.

In Fig. 2.3b the rotation orbits of the two agitators overlap, so that the mixing operation is more efficient and the mixing time is shortened. The rotational speeds are selected so that there is no interference between the two agitators. The speed ratio is usually 1:1 or 1:2. For the ratio 1:2, this results in the faster agitator pursuing the slower one so that the dough between the two agitators is fully stretched, folded, and kneaded very quickly.

2.2.2   Mixing time
The output of the mixer is dependent on the formulation and the degree of automation of the ingredient feeding process. Mixing times, including ingredient feeding, mixing, and discharging, usually range from less than 7 minutes to more than 30 minutes, depending on the recipe and the required final consistency of the dough.

The developed doughs are normally mixed for a predetermined time by means of Brabender instruments. For short and soft doughs, the mixing time is determined by trial-and-error based on the operator's judgement.

Some modern mixers are equipped with a thermometer or thermocouple to control the operation. Temperature control has two uses. One is in mixing temperature-sensitive doughs for which an intensive blending action is not required. [1] The other is for doughs requiring intensive kneading and stretching. As the dough temperature reaches the preset value, the machine stops, the desired energy having been put into the dough. Further mixing will make the dough lose its desirable rheological properties, which will make it difficult to manipulate in the next stage.

Some modern machines have a sprag inserted into the centre part of the mixing shaft (Fig. 2.5), while others have a strong 'horn' which is fixed on the centre part of the mixing bowl and projects into the dough mass. [4] Inside the sprag or 'horn' is a temperature sensor which is connected to the control system. In either configuration,

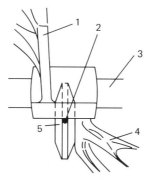

Fig. 2.5   Location of mixing arm sprag sensor (Courtesy of Vicars Group Ltd) (1) Mixing
arm, (2) Temperature sensor, (3) Centre shaft, (4) Mixing arm, (5) Sprag.

the mixing arms would be divided into two parts, right and left side. Fig. 2.5 shows the
relative location of the mixing arm-sprag-sensor of a modern mixer. Table 2.3 shows
specifications of this type of mixer.

2.2.3   Control
Modern mixers are equipped with an advanced control system. On the facia control,
there are automatic time controls for electronically linked low and high speed process
timers for the duration and intensity of the mixing action, automatic dough
development control by accurate sensing of dough temperature, computer control of
the ingredient feed and mixing process, and start and stop buttons. For safety, the
operator must use two hands to operate two buttons simultaneously to discharge the
completed dough. As these buttons are released, a pneumatic friction brake automatically stops the mixing arm. Modern mixers are also fitted with front and rear safety
guards, and a safety tub lock which ensures that either a tub or chute is in position
(APV Baker).

2.2.4   Principle of dough mixing
From the standpoint of food technology, doughs may fall into the following
categories: hard dough for biscuits, short and soft dough for cookies, and sponge

Table 2.3   Specifications of some large mixers with centre shaft (Courtesy of Vicars
Group Ltd)

| Model | Dough capacity (kg) | | Mixing speed (rev/min) | | Main motor | |
	Hard dough	Soft dough	Single	Dual	Horse power	kW
1	300	350	90	90/45	65	49
2	455	545	75	75/37	100	75
3	636	727	75	75/37	150	112
4	900	1000	60	60/30	150	112

dough for breads. But from the mechanical standpoint, doughs are classified into extensible doughs and friable doughs (or mixtures). The mixing principle of extensible doughs will be mainly discussed since it presents the more complex problem.

The extensible doughs include the unleavened doughs for Chinese snacks, which will be discussed in following chapters, and the leavened doughs for biscuits, bread, and rolls.

Dough mixing, as mentioned previously, has two stages. Firstly, the ingredients are blended. During this period, the most significant function of mixing is the hydration of the flour (more correctly, the gluten proteins). This occurs very quickly and is carried out at low speed with low power consumption. Secondly, the gluten network structure is gradually developed, which results in a gradually increased torque to the mixing arms with much more power and mixing time needed than in the first stage. To shorten the mixing period, the intensive mixing technique is commonly used in the bakery field. The speed for gluten development is double that for the first stage, and power consumption is greater. The rated power of the main motor of the mixer is calculated and chosen according to the greatest torque from the developed dough to the mixing spindle.

Fig. 2.6 illustrates the mixing principle. The mixing arms 1, 2, 3, and 4 are mounted on the figure-S support plates. The dough development process is as follows.

(a)   The clockwise rotating arms 1 and 2 lift and carry the dough from the front to the back of the mixer bowl.
(b)   With the limited clearance between the mixer walls and the rotating arm orbit, the dough mass is beaten and stretched by arms 1 and 2.

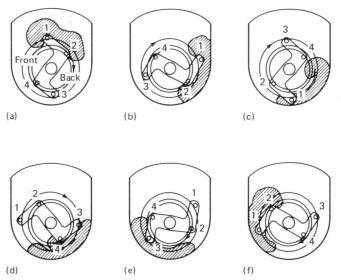

Fig. 2.6   Dough mixing principle (Courtesy of Hai Yi Tao (Height) Inc., Japan).

(c)   The dough is further stretched by arm 1 and folded by arm 4 as arm 2 rotates away from the dough.

(d)   As arm 1 leaves the dough, arms 3 and 4 arrive at the bottom of the bowl so that the stretched dough is pressed folded, and kneaded again by arms 3 and 4 at the bowl bottom.

(e)   As the arms rotate, the dough is stretched and pulled by arm 3 from the bottom toward the back of the mixing bowl.

(f)   As arm 3 leaves, arms 1 and 2 engage with the dough at the back of the bowl. The limited clearance of the mixer walls by the rotating orbit of arm 1 and 2 causes the dough to be stretched again and further carried to the bowl front where the next mixing cycle begins.

The explanation of this process is lengthy, but the operation is very fast, so that in a very short period the dough mass is repeatedly stretched, folded, and kneaded in the same direction. This results in oriented gluten fibres and dough development. To a large extent, the mixing response of flour is dependent on the amount and the quality of the gluten. But the correct selection of mixing agitators is also very important for dough quality. A suitable choice will lead to the successful formation of a continuous gluten network, which is necessary for the dough to have the maximum gas holding capacity upon which depends the quality of the finished products, especially the specific volume in bread production.

2.2.5 Transmission

The transmission system of the horizontal mixers is generally of simple construction. It comprises a large electric motor referred to as the main motor for mixing drive, and a small reversible motor for bowl tilting to discharge the dough mass and for restoring the mixing position of the bowl. Since the mixing speed and bowl tilting speed are always much lower than the motor speed, the reduction ratio used is very large, and a reducing gear should be used, which is either a reducing worm-gear or a planetary gear reducer.

The worm-gear has a lower transmission efficiency, but is cheaper and easier to use, so that it is used in most bowl-tilting mechanisms and some smaller capacity mixing transmission systems (as shown in Fig. 2.7).

The planetary gear reducer has a compact construction, higher transmission efficiency, and lower power consumption, but costs more than the worm-gear. It is widely used in large capacity mixing transmission system as shown in Fig. 2.8 in which the motor and reducer are assembled together.

In Fig. 2.7, the rotation is transmitted from main motor (1) through pulleys (2) and (4) to worm shaft II on which worm (5) is engaged with worm gear (6). The mixing arms are driven by shaft III which is the mixing spindle actuated by worm gear (6). For bowl tilting, the rotation is transmitted from motor (8), through pulleys (9) and (10), worm (11) to worm gear (12) which is fixed on the side wall of the mixing bowl so that the bowl can rotate with worm gear (12) to tilt for dough discharging. In reverse it restores the mixing position. This transmission expression is shown in Scheme 2.1.

$$\text{Motor 1} \rightarrow \text{Shaft I} \rightarrow \frac{\text{Pulley 2}}{\text{Pulley 3}} \rightarrow \text{Shaft II} \rightarrow \frac{\text{Worm 5}}{\text{Worm gear 6}}$$
$$\rightarrow \text{Shaft III(mixing spindle mixing arms 7)}$$

$$\text{Motor 8} \rightarrow \text{Shaft V} \rightarrow \frac{\text{Pulley 9}}{\text{Pulley 10}} \rightarrow \text{Shaft IV} \rightarrow \frac{\text{Worm 11}}{\text{Worm gear 12}}$$
$$\rightarrow \text{Bowl tilting and restoring}$$

In Fig. 2.8, the bowl-tilting mechanism is the same as in Fig. 2.7, but the mixing arms are driven by motor (1) through shafts I and II, chain wheel (3), (4), and planetary reducer (2). This mixing transmission expression is as follows.

$$\text{Motor 1} \rightarrow \text{Planetary reducer 2} \rightarrow \text{Shaft I} \rightarrow \frac{\text{Chain wheel 3}}{\text{Chain wheel 4}}$$
$$\rightarrow \text{Shaft II (mixing spindle to arms 5)}$$

Scheme 2.1.   Transmission scheme for horizontal mixer.

Since the horizontal mixers are very easy to run even for an unskilled operator, it is not necessary to talk about its 'operation', as no mechanical knowledge is needed.

2.3   VERTICAL MIXERS

Vertical mixers have a vertical mixing spindle. Fig. 2.9 is the front view of a typical mixer.

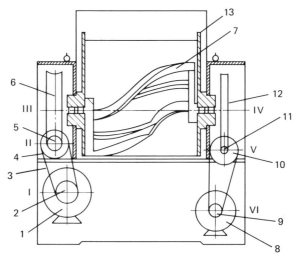

Fig. 2.7 'Shaftless' mixer with figure-S mixing arms (Courtesy of Xin Xiang County Food Machinery Factory, China) (I) to (V) Shafts. (1) Main motor, (2) Pulley, (3) Belt, (4) Pulley, (5) Worm, (6) Worm gear, (7) Arm, (8) Bowl-tilting motor, (9), (10) Pulleys, (11) Worm, (12) Worm gear, (13) Mixing bowl.

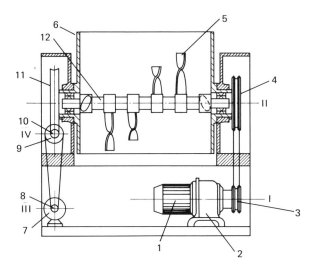

Fig. 2.8   Horizontal mixer with centre shaft (Courtesy of Shang Hai Food Machinery Factory, China) (I), (III), (IV) Shafts, (II) Mixing spindle. (1) Main motor, (2) Planetary gear reducer, (3), (4) Chain wheels, (5) Mixing arm, (6) Double hopper, (7), Reversible motor, (8), (9) Pulleys, (10) Worm, (11) Worm gear, (12) Centre shaft.

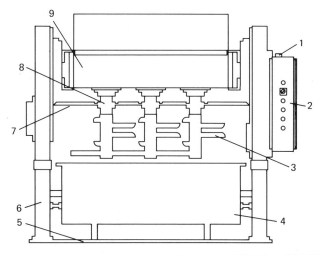

Fig. 2.9   Front view of a modern vertical mixer (Courtesy of APV Baker, UK) (1) Electrical connection, (2) Facia control, (3) Mixing arm, (4) Dough tub, (5) Base plate, (6) Post, (7) Rail, (8) Mixing spindle, (9) Bridge.

2.3.1 Construction

A vertical mixer has a portable dough tub (bowl) which is charged with all the dough ingredients before being placed under the mixing head. The mixing head comprises mixing spindles and arms which are driven by a main motor through a reduction gearbox and a final oil-less chain drive or other well-sealed mechanism to prevent the contamination of the mixing spindle and arm.

Depending on the designer, the mixing operation can be carried out either by means of lowering the mixing head or raising the tub against the head. In either case, another motor is needed, which is reversible and is referred to as the secondary motor. As soon as the mixing has been properly done, the main motor is automatically shut off by the controlling system, and the secondary motor reverses, either for raising the mixing head or for lowering the dough tub. Table 2.4 shows specifications of some vertical mixers.

The main frame of the machine is either of iron construction or of unitary construction of heavy steel plate. The thick base plate is designed for carrying the robust corner posts of the mixer, and it is mounted at floor level to facilitate tub movements as well as for stable mixing and easy cleaning. For food hygiene and operational safety, the transmission mechanisms for the mixing spindles are housed in the bridge (Fig. 2.9), while the Vee belt drive motors are fully guarded at the rear of the machine, clear of the floor.

Vertical mixers are commonly equipped with two or three spindles, each of which is fitted with three to six shaped arms (blades) arranged in a broken spiral. The bottom arm is designed to operate in close proximity to the bottom of the dough tub. As shown in Fig. 2.9, the three five-armed agitators overlap and pass each other in opposite directions, giving an efficient mixing action. The spindles are usually made of high tensile steel with top splined driving ends, whose lower ends are located in the cross rail by self-aligning ball races.

The mixing arm of the vertical mixers is commonly shaped in a simple oar-like form, but in various sizes to match different capacities. They normally are forged, to

Table 2.4   Specifications of some vertical mixers (Courtesy of APV Baker, UK)

Model	1	2
Mixing mode	Dough tub lifted	Spindle lowered
Batch capacity	475 kg for very hard dough 550 kg for softer dough	635 kg
Mixing spindle	2	3
Mixing speed	25 rev/min	25 rev/min
Main motor	22.5 kW	22 kW
Secondary motor	reversible, 2.2 kW	reversible, 2.2 kW
Dimensions	2132 × 1380 × 2928 (mm)	3099 × 1448 × 2820 (mm)
Net weight	5840 kg	

ensure strength, and are coated with stain resistant materials for easy cleaning and food hygiene.

The mixing speed of the machine is commonly in a single-speed mode, about 25 rev/min, which is lower than that of horizontal mixers.

This group of mixers can be used, if necessary, for almost any doughs, but their principle advantages are for doughs which are to be fermented (such as bread or soda biscuit doughs) because the sponge and dough do not have to be transferred in and out of the mixing tub between various stages. Because of this, a vertical mixer is always provided with more than one tub to maintain the dough supply. For example, one is being filled with flour and other ingredients, one is mixing, one is being used for dough fermentation, and perhaps one more is being emptied and is waiting for recharging. Since the tub can move by means of four ferrous tread castors mounted to its underside, it can be charged at the filling centre so that the separate pipes and weighing systems needed for individual horizontal mixers are omitted. However, it also brings some disadvantages, such as the space necessary and labour costs.

Tubs, which correspond to the mixing bowls of the horizontal mixers, are flat at the bottom with figure of eight shaped sidewalls for two-spindle mixers, or with O-shaped walls for three-spindle mixers. The tub body is commonly fabricated in heavy gauge mild steel. A substantial lifting frame welded to the tub underside is needed for the dough discharging operation in which tines on the tub engage with the lifting and tilting mechanism. A lift-off lid allows manual emptying of the tub.

2.3.2 Transmission
As shown in Fig. 2.10, the rotation of main motor (1) is transmitted through a multi-rope Vee belt and pulley (2), (3) to shaft II, by which worms (4) are actuated. The mixing spindles (II), (IV) are driven by worm-gears (5) which engages with worm (4).

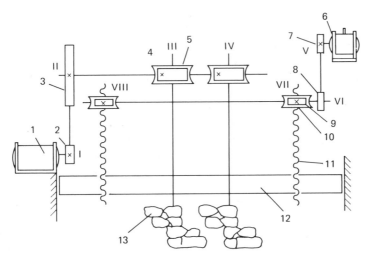

Fig. 2.10   Transmission system of a typical vertical mixer (I), (II), (V), (VI), Shafts, (III), (IV) Mixing spindles (VII), (VIII), Lifting and lowering screws. (1) Main motor, (2), (3) Pulleys, (4) Worm, (5) Worm-gear, (6) Secondary motor, (7), (8) Pulleys, (9) Worm, (10) Worm-gear, (11) Screw, (12) Rail with screw jack, (13) Mixing arm.

The worm-gearing (4) and (5) change the direction of motion and transmit to the mixing spindle III and IV so that the mixing arms are made to rotate in the same direction.

In this machine, the mixing is carried out by means of the mixing spindle lowered into the underlying tub. For this, the secondary motor (6) transmits its rotation through pulleys (7) and (8) as well as another two sets of worm gearing mechanisms to screw (11) which is linked to shaft (VII) and (VIII). The rotation of screws (11) makes the screw jacks on the horizontal rail (12) to move down so that the mixing arms (13) are lowered into the working position along the side guides. Reversal of the secondary motor will drive the mixing spindles upwards to the upper position to allow the tub to be removed. The splined driving ends of the spindles make the up and down motion possible.

Attention should be paid to the lubrication of the worm gearing mechanisms and the self-aligning bearings; no oil or grease leakage is permitted.

The transmission expression of this machine is shown in Scheme 2.2.

$$\text{Motor } 1 \rightarrow \frac{\text{Pulley } 2}{\text{Pulley } 3} \rightarrow \text{Shaft II} \rightarrow \frac{\text{Worm } 4}{\text{Worm gear } 5} \rightarrow \text{Spindle III} \rightarrow \text{Mixing arm } 13$$

$$\text{Spindle IV}$$

$$\text{Motor } 6 \rightarrow \frac{\text{Pulley } 7}{\text{Pulley } 8} \rightarrow \text{Shaft VI} \rightarrow \frac{\text{Worm } 9}{\text{Worm gear } 10} \rightarrow \text{Shaft VII} \rightarrow \text{Screw } 11 \text{ (to screw}$$

$$\text{VIII} \quad \text{jack 12 for}$$
$$\text{spindle up,}$$
$$\text{down)}$$

Scheme 2.2.   Transmission scheme for a vertical mixer.

2.3.3   Working process and control

Before operation, the tub enters directly from the front of the mixer and is locked onto the baseplate by self-centring four-point locators operated by a handwheel (Vicars Group Ltd).

To start the machine, the button on the facia mounted on the side of the machine (Fig. 2.9) is pushed to lower the mixing spindles. As soon as the mixing arms arrive at their working position in the tub, the lowering motion is stopped immediately by a travel switch. The mixing start button is then pushed. Modern machines are generally designed with a safety circuit which ensures that the mixing arms can rotate only when the tub is in its correct position. The rotating arms sweep the entire volume of the dough tub. Since the shaped arms are fixed on the spindle at various angles, the dough in the tub is forced upward or downward, and is gently blended, sheared, and rolled over and over again until all the ingredients are mixed thoroughly and form a uniform dough mass. The amount of gluten development is, to some extent, influenced by the configurations of the mixing arm. Since the arms of this group of mixers often provide a more or less shearing and cutting action to the dough, it is unfavourable to gluten development but suitable for short and soft dough mixing. For bread doughs, the arms should be designed with the least possible cutting action but greater stretching and kneading effect.

Mixing time is dependent on the type of dough: for semi-sweet dough about 30 to 60 minutes, for soft dough 15 to 30 minutes, and for soda cracker dough and sponge doughs about 4 minutes (at a mixing speed of 25 rev/min, batch capacity 635 kg, for a three-spindle type mixer, APV Baker). The mixing time can be controlled directly by pushing the 'mixing stop' button on the facia panel or automatically by a pre-set timer. After the mixing arms stop completely, the secondary motor is started manually or automatically to raise the mixing spindles to their upper position to allow the dough tub to be removed.

As previously mentioned, there is another type of vertical mixer in which the tub is raised for mixing rather than lowering the mixing spindles. In this design the secondary motor is in charge of the dough tub lifting and lowering. A sensor is incorporated to ensure that the tub is in position before the tub lift will operate. The mixing arms will stop automatically if the tub is lowered. To lower the tub, the 'tub lower' button must be held in until the tub is fully down. If the button is released, the tub movement will stop automatically.

2.4 COMPARISON OF HORIZONTAL AND VERTICAL MIXERS

The two groups of mixers have their advantages and disadvantages. Taking all aspects into consideration before buying new mixing equipment can make your investment beneficial.

Comparing the vertical mixers with the horizontal ones, the characteristics are as follows.

(1) From the mechanical standpoint, the mixing agitators of horizontal mixers have bearings at each end, giving better working conditions than the vertical mixers. The latter have mixing spindle bearings only at their driving ends, and their mixing ends are cantilevered, which results in great bending torque and necessitates more sophisticated calculations in the design of the mixing spindles.

(2) The horizontal mixers provide various mixing arms with different configurations which allow a wide range of doughs to be mixed more effectively than in the vertical mixers, especially for developing the extensible doughs, since some mixing arms are especially designed with blending, stretching, and kneading actions rather than shearing and cutting effects which are often inevitable in vertical mixers.

(3) The construction of horizontal mixers allows continuous control of the dough temperature since the temperature sensor and coolant circuiting jacket are easier to incorporate than in the vertical mixers in which the tub has to be taken out at the end of each cycle, making temperature control more difficult.

(4) The horizontal mixers allow the ingredients to be fed into the mixing bowl at any mixing stage, with the mixing arms rotating. This is especially favourable for cookie doughs. Vertical mixers are always fed before mixing.

(5) Horizontal mixers need less space, and are easier to include in an automatic production line, since tilting the bowl discharges the dough into an underlying hopper for feeding to the next production stage. The tubs of vertical mixers have always to be detached and taken to another place where the dough is discharged by an extra lifting and tilting device. This is time- and labour-consuming work.

(6) The tubs of the vertical mixers can be taken away to the filling centre. Consequently, their weighing and feeding systems are simplified and easier to control. Furthermore, pipelines are shortened, so that power consumption and mechanical maintenance are reduced.

(7) The detached tub of the vertical mixers allows it to be double-used. That is, the tub can be used as a part of the mixer for mixing; afterwards, it can be used as a dough fermentation vessel so that the dough discharge operation is omitted, which results in time- and labour-saving and a saving in extra fermentation equipment.

(8) Generally speaking, vertical mixers are low-speed mixers providing a gentle mixing action which results in a lower dough temperature during mixing. This is very beneficial to the doughs which must be kept cool for the next stage of operation and for doughs containing easily damaged ingredients. Horizontal mixers often work at high speed for dough development, which leads to a vigorous dough mass movement, extensive stretching, rolling, and kneading, accompanied by a violent friction between the dough and the bowl. As a result, the dough temperature will rise much higher than in vertical mixers, so that large horizontal mixers are often equipped with a coolant jacket and are sometimes fed with chilled ingredients among which chilled water is the most convenient.

(9) To some extent, vertical mixers are easier to manufacture, easier to install and to clean than the horizontal mixers in which the mixing actions often give considerable vibration, especially when installed above the ground floor, so that specially constructed bases are needed for large horizontal mixers, which are located and fixed by means of fang bolts.

(10) It is often recommended that the two types of mixers should be used in a production line where the dough needs more than one mixing stage. The horizontal mixer is for the first mixing, and the finished dough is discharged into the vertical mixer's tub for fermentation and later to the vertical mixer for the second stage of mixing. In this way, their advantages can be fully utilised, and some disadvantages can be balanced out.

2.5   OTHER TYPES OF MIXERS

2.5.1   Rotatable-bowl mixers

The mixers already described have static bowls or tubs during mixing. There is another group of mixers in which the dough bowls slowly revolve as the mixing arms either rotate or reciprocate so that the dough mass is under a compound motion which gives efficient dough mixing.

This type of mixer is designed with a smaller batch capacity, up to 250 kg of dough.

Table 2.5 Specifications of some mixers with rotary bowl (Courtesy of Baker Perkins Bakery Ltd)

Model	1	2	3
Max. flour/mix (kg)	75	100	150
Dough (kg)	120	160	250
Head motor (kW)	3.0/5.0	5.0/9.0	6.5/11.0
Bowl motor (kW)	1.1	1.1	1.5
Dimensions (mm)	1350 × 910 × 1260	1470 × 960 × 1280	1730 × 1110 × 1400
Net weight (kg)	710	845	1350

It is suitable for fermented dough, straight dough, sponge dough, and pastry. Table 2.5 gives the specification of mixers of this type.

In this group, the bowls are made of highly polished stainless steel to assist cleaning and hygiene.

Some machines are equipped with two electric motors to drive the dough mixing bowl and the mixing arm respectively, while other machines are fitted with only one motor to drive both the bowl and the mixing arm at the same time through corresponding transmission mechanisms. In any case, the bowl always receives the torque and motion from the bottom drive, while the mixing arm is always actuated through its upper end by the driving spindle.

The mode of motion of the mixing arms is classified into rotating and reciprocating type.

In the rotating mode, the mixing arms (agitators) are usually designed with spiral configurations. One mixer is generally equipped with one spiral mixing arm, which is either vertical or inclined to the bowl bottom. As the bowl is slowly revolving, the arm rotates in its fixed place. The relative motion ensures that the dough is mixed thoroughly. The spiral configurations of the arms provide efficient blending, rolling, and kneading actions without shearing and a cutting effect which is favourable to the development of gluten. Fig. 2.11 is a schematic transmission diagram of a typical rotary-bowl mixer with a spiral mixing arm. For this machine, one electric motor is fitted, from which the rotation is transmitted in two ways: through pulleys (2), (3) and worm gearing (4), (5) to shaft (III) which causes the bowl (10) to revolve slowly; or through pulleys (2), (6) and worm gearing (7), (8) to the mixing spindle (V) which actuates the mixing arm (9) to rotate at a higher speed than the dough bowl. The bowl speed is about 10 rev/min, and the mixing arm speed is under 50 rev/min.

In the reciprocating mode, a pair of mixing arms is often fitted. They are commonly designed to travel through intersecting elliptical paths in a shallow, slowly revolving bowl and to impart to the dough a gentle kneading, stretching, lifting and folding action [2]. For the relatively slow rate of energy input and the gentle manipulation, the dough temperature is barely increased during mixing. The machine is especially suitable for doughs containing nuts or raisins and for pie and pastry doughs. However, this kind of machine is not commonly manufactured or used.

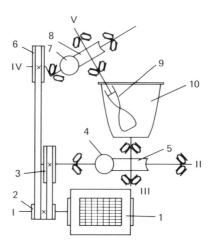

Fig. 2.11 Transmission system of a typical mixer with a rotary bowl (Courtesy of Shang Hai Food (Yin Shi) Machines and Tool Factory, China) (I), (II), (IV) Shafts (III) Bowl rotating spindle, (V) Mixing spindle (1) Motor, (2), (3) Pulleys, (4) Worm, (5) Worm-gear, (6) Pulley, (7) Worm, (8) Worm-gear, (9) Mixing arm, (10) Bowl.

2.5.2 Biplex mixers

This type of mixer uses the patented Biplex process, which was introduced in the 1970s by the British APV Baker firm. The unique feature of this process is that two separate stages are performed by a spiral mixing arm in the same mixing bowl to produce optimum dough development which is ready for immediate dividing without the conventional fermentation stages.

Stage 1, as shown in Fig. 2.12, is a mode of planetary action which progressively works the dough, incorporates and hydrates the ingredients, and draws air into the mixture. In stage 2, as the planetary rotation stops, a higher intensive single rotation is produced to develop the dough in a very short period with a minimum work input. If the dough recipe is for the Chorleywood bread process, it would be 'mechanically' fermented with a good build-up of gas cell structure so that it can be directly transferred to a divider. In fact, the Biplex process produces fully developed doughs with a lower energy input of only 7–8 Wh/kg than the Chorleywood process standard of 11 Wh/kg, and also with lower yeast and improver or higher product volume. The mixing time is based on only 3 minutes, while the ratio of flour to water is 100:60.

The machine is available as a basic mixer or with a host of options, including a weighing system, hoist, etc. and various levels of controls for either semi-automatic plant or fully automatic production lines. Table 2.6 shows some basic specifications of the Biplex mixers.

2.5.3 Continuous mixers

Different from the previous mentioned mixers, the continuous mixers are in a continuous mode, continuously feeding, mixing, and extruding the finished dough in a continuous stream, rather than in batches.

Stage 1

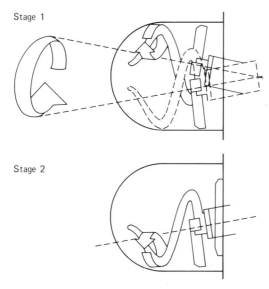

Stage 2

Fig. 2.12   Schematic diagram of the Biplex mixer working process (Courtesy of APV Baker,
UK).

Table 2.6   Specifications of some Biplex mixers (Courtesy of APV Baker, UK)

Model	1	2	3	4
Maximum dough output (kg/h)	3000	4000	4500	6600
Max. dough batch size (kg)	250	333	365	660
Max. flour capacity (kg)	155	205	225	400
Main motor (kW)	55	75	75	75
Total connected power (kW)	64	86	86	95

The mixing functions of most continuous mixers are performed by one or two rotors in a cylindrical casing, which are used as stators. The rotor and stator are carefully designed to provide mixing and conveying actions for the dough to be developed and extruded out of the pressure control flap in a continuous ribbon of dough.

For different doughs, not only the rotor but also the stator should be changed. That means a complete change for the mixer which is normally used on single-purpose, high-output lines or on similar products. [4] Furthermore, continuous mixers need much more prior study, and they cost much more for their weighing and controlling systems as well as for installation and maintenance than batched horizontal and vertical mixers. Continuous mixers are not used very widely today. This group of mixers have only a brief introduction in this book.

Part II

Chinese snack-making machines

3

Steamed bun manufacture

Chinese food occupies a bright and colourful place in the blossomy flower garden of world food. It is a part of China's ancient civilization and long history. In recent years, it has attracted more attention, and is becoming more and more popular. To taste Chinese food is a pleasure of life for the visitors to China. The author would like to introduce some typical Chinese snacks, which are easy to make by hand. If there is no Chinese restaurant nearby, why not try to do it yourself? As you taste the delicious food, you will get much pleasure from it.

3.1 INTRODUCTION

In China, even in Southeast Asia, the 'steamed bun' is as important as baked bread in the West. The bun is also referred to as 'steamed bread', which is also a fermented food. But its technology is not so time-consuming and complicated as the baked bread process. The steamed bun is called 'Man Tou' or 'Mo' in Chinese. It can be used for sandwiches, toast, and other types of food. The 'Xi An Pao Mo' is very famous, and is popular with the Chinese and with many foreign visitors. 'Xi An' is the name of Shannxi province capital which was the capital of the Tang Dynasty in a period of great prosperity in China. 'Pao' means to soak the pieces of bun (Mo) in a very tasty and special soup.

The dough (with the ratio of flour to water 2:1) is traditionally fermented by means of a small piece of leavening dough. Nowadays, some yeast is employed to shorten the fermentation time from some hours to about forty minutes. When made in this way, the finished buns have a slightly different flavour from the traditional product, but they are very soft and tasty.

Doughs fermented in the traditional way must first be neutralized by sodium carbonate or sodium bicarbonate, and then left to stand for a while before forming into buns either by hand in the domestic kitchen or by special machinery in the restaurant kitchen. The well-rounded dough pieces will be put into a steam cooker. Twenty minutes later, the dough balls are well-steamed into white, smooth, appetizing buns with a volume two to three times that of the original dough piece.

Since the dough for bread contains more water than that for steamed buns, the latter is much harder than the former. Furthermore, bread dough is a complex mixture of flour, water, sugar, syrup, eggs, butter, milk, yeast, and so on, while bun dough is only the mixture of flour and water. So the latter is always much more elastic but much less plastic than the former. That is why bread rounding machines cannot be used in the steamed bun process.

The newly developed steamed bun maker is a simplified but very effective machine which combines the dough feeding, sizing, and shaping to produce dough balls of high quality. The steamed buns shaped by the bun maker are the only type of snack whose quality and taste are better than those of the traditionally hand-made buns.

Fig. 3.1 is a schematic diagram of a typical steamed bun maker, and Table 3.1 shows the general specifications of some bun makers.

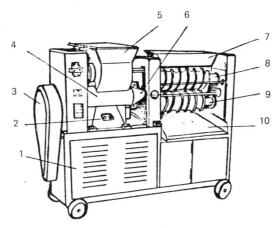

Fig. 3.1 Typical steamed bun maker (Courtesy of Xi An Food (Yin Shi) Machinery Factory, China) (1) Motor (inside) (2) Switch box, (3) Belt-wheel assembly, (4) Dough chamber, (5) Dough hopper, (6) Dough outlet, (7) Flour dusting device, (8) Upper shaping roll, (9) Lower shaping roll, (10) Chute.

Table 3.1   Specifications of typical steamed bun makers

No.	1	2	3
Capacity (flour weight kg/h)	360	360	360
Scaling (flour g/piece)	100	100	100
Motor	3 kW 3-phase 380 V	3 kW 220 V	3 kW 3-phase 380 V
Dimensions (mm)	1420 × 500 × 1100	1300 × 400 × 900	1500 × 640 × 1200
Net weight (kg)	300	260	500

3.2 CONSTRUCTION

The machine consists of dough feeding, sizing, and forming assemblies, an electric motor, control, and transmission system.

3.2.1   Dough feed assembly (Fig. 3.2)
This assembly comprises a hopper, a dough pushing device which consists of a driven shaft, and two spiral blades to press the dough into the slots of the dough screw conveyor. The hopper is fixed to the feed chamber on the inside of which is the screw which consists of two sections (introduction and working sections), while the screw sleeve is mounted on the right side of the dough chamber. The left part of the screw in the dough chamber is larger in diameter than the right part, which is conical. The cast feed screw is coated with Teflon to avoid sticking. Over many years, the parameters for the dough feed assembly have become standardized. Table 3.2 shows the recommended data.

3.2.2   Forming assembly
Fig. 3.3 shows the forming assembly consisting of a pair of shaping rolls, a flour dusting device by which the flour is well-distributed on the working area of the shaping rolls, and a chute along which the rounded dough pieces roll down. Under the chute is either a web or a pan to receive the dough pieces; this leads to a steam cooker.

   The left end of the shaping rolls has a blade (edge) to cut the dough once at each rotation of the shaping roll. The rolls are very important for the shaping of the dough

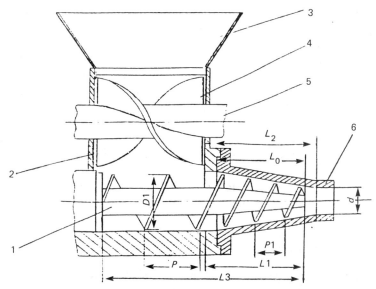

Fig. 3.2   Dough feed assembly (Courtesy of Harbin Food and Service Machinery Research Institute, China) (1) Dough feed screw, (2) Dough chamber, (3) Dough hopper, (4) Plucking blade, (5) Shaft, (6) Screw sleeve. See Table 3.2 for dimensions.

Table 3.2   Recommended parameters of the dough feed screw† (Courtesy of
Harbin Food and Service Machinery Research Institute, Harbin, China)

Bun scaling (flour g/piece)	d (mm)	D1 (mm)	P (mm)	L0 (mm)	P 1 (mm)	L1 (mm)	L3 (mm)	L2 (mm)
50	40	85	50	125	50	140	300	140
		106	100	160	50	180	375	260
100	45	150	160	100	67	140	355	140

†Where *d* is the diameter of dough outlet, *D1* is the screw diameter of the introduction section, *P* is the screw pitch of the introduction section, *L0* is the length of the cone, *P1* is the pitch of the working section, *L1* is the length of the working section, *L3* is the length of screw, *L2* is the length of the compression section.

pieces. If the rolls are too long, this would allow the dough to decrease in the elasticity which is wanted for a good product. Short rolls would produce a dough piece uncompacted and uneven in structure, resulting in a rough skin and undesirable taste. The buns should be bright and white, with a very thin skin with a layered structure, and should be fluffy and sweet-smelling.

To obtain satisfactory products, much work has been done. Both theoretical and practical parameters are summarized in Table 3.3. The shape of the roll's curved surface is dependent on the size and the shape of the dough piece formed. Since the steamed buns are usually in two sizes, 100 g and 50 g of flour weight, the weight of each dough piece would be about 150 g and 75 g respectively: as the ratio of flour to water in this type of dough is about 2:1. Experimental work shows that at 150 g well-rounded dough piece is about 134 cm^3 in volume and 64 mm in diameter, which means that the circle formed by the two rolls should be about 64 mm in diameter. Thus the radius of the hemispherical curve of each shaping roll should be about 32 mm. However, there is always a gap between the two parallel rolls, so that the practical radius of the mould curve is about 28 mm to meet the preset volume of the finished dough pieces which is about 134 cm^3. Since the rounding process is to change the dough from a loose to a compacted structure, and from larger to smaller volume, the moulding curve depth is shallower from the left to the right (that is, from the beginning to the end of the rounding process).

3.3   FORMING PRINCIPLE

The shaping process of the steamed bun maker includes three stages: feeding, sizing, and rounding. As shown in Fig. 3.2, the dough pushed by the plucking blade in the hopper goes into the slots of the dough feed screw in the chamber, which introduces the dough into the feeding assembly. That is why this part of the feed screw is referred

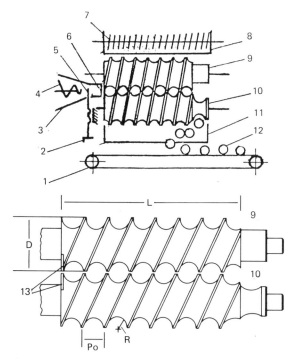

Fig. 3.3   Forming assembly and shaping roll. (1) Receiving conveyor, (2) Adjusting hand-wheel, (3) Screw sleeve, (4) Dough Screw, (5) Adjusting Valve, (6) Dough outlet, (7) Brush, (8) Flour hopper, (9) Upper shaping roll, (10) Lower shaping roll, (11) Chute, (12) Dough ball, (13) Cutting edge. See Table 3.3 for dimensions.

Table 3.3   Recommended parameters of the shaping roll† (Courtesy of Harbin Food and Service Machinery Institute, Harbin, China)

Bun sizing (flour weight g/piece)	$P0$ (mm)	D (mm)	L (mm)	L/D	n (rev/min)	P (kW)	R (mm)
50	52	125	450	3.6	66	2.2	22
					60	3.0	
100	64	150	450	3.0	100	4.5	28
					50	3.0	
			500	3.3	100	4.5	
			710	4.7	60	3.0	
					100	4.5	

†Where $P0$ is the pitch of the shaping roll, D is the roll diameter, L is the effective length of the shaping roll, L/D is the ratio of the effective length to the diameter of the shaping roll, n is the speed of the shaping roll, P is the maximum power of the motor, and R is the radius of the moulding curve of the shaping section of the roll.

to as the introduction section, by which the dough is further conveyed into the conical part of the feed screw. In this part, the dough is compressed little by little, since the screw diameter becomes smaller and smaller (conical). So the dough structure becomes denser, which is helpful for the later sizing and rounding. For this reason, the conical section of the dough feed screw is referred to as the working section or compression section. From the outlet of the screw sleeve, the dough is pushed into the forming assembly and engaged by the face-to-face revolving rolls. Here, the dough is cut into pieces by the blades on the left end of the two shaping rolls. In the moulding slots, the pushing force from the left-side wall of the slot makes the dough piece move forward to the right, while the coupling of the forces from the moulding curve of the rotating rolls makes the dough roll, and rotating along in a rectilinear motion between the two parallel rolls. From the left to the right, under the compound forces, the irregularly shaped dough pieces are rounded into compacted smooth balls with a uniform texture in a compound motion. The balls move further and fall down onto the chute along which they roll to the receiving web which leads to the steam cooker.

3.4   TRANSMISSION SYSTEM

Fig. 3.4 is a transmission diagram of a typical steamed bun maker. The machine is driven by an electric motor (3 kW, 3-phase, 50 Hz, 380 V) through a one-stage belt transmission mechanism and a series of gear trains.

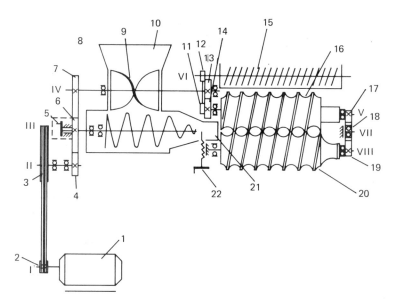

Fig. 3.4   Transmission system of a typical steam bun maker (Courtesy of Xi An Food (Yin Shi) Machinery Factory, China) (I)–(VIII) Shafts (1) Motor, (2), (3) Pulleys, (4) Gear wheel, (5) Handwheel, (6), (7), Gear wheels, (8) Dough feed screw, (9) Plucking blade, (10) Hopper, (11), (12) Pulleys, (13), (14) Gear wheels, (15) Flour dusting brush, (16) Upper shaping roll, (17), (18), (19) Gear wheels, (20) Lower shaping roll, (21) Screw sleeve outlet, (22) Handwheel.

From motor (1), the rotation is transmitted through pulleys (2) and (3), shaft (II), gear wheels (4) and (6) to shaft (III) which actuates the dough feed screw to convey the dough into the forming assembly.

From shaft (III), through gear wheels (6) and (7), the rotation is further transmitted to shaft (IV) which causes the plucking blade (9) to rotate so that the dough is pushed downward into the dough chamber.

From shaft (IV), the rotation is further transmitted through gear wheels (13) and (14) to shaft (V), which drives the upper shaping roll (16), while the lower shaping roll (20) is actuated by shaft (VIII) which receives the motion from shaft (V) through gear train (17, 18, 19). The idler gear (18) makes it possible for the rotation to be transmitted to the lower shaping roll and at the same time the rotation direction to be changed to allow the upper and lower rolls to rotate in an opposite direction (face-to-face), which is the prerequisite of the rounding process.

Also from shaft (V), the rotation is transmitted through a pair of small rope pulleys (11) and (12) to shaft (VI) by which the brush rotates through the bottom sieve of the flour hopper to distribute the flour on the shaping roll and the dough piece to prevent sticking and increase friction, which is helpful for the rounding process.

The transmission route is shown in Scheme 3.1.

$$\text{Motor 1} \rightarrow \text{Shaft I} \rightarrow \frac{\text{Pulley 2}}{\text{Pulley 3}} \rightarrow \text{Shaft II} \rightarrow \frac{\text{Gear wheel 4}}{\text{Gear wheet 6}} \rightarrow \text{Shaft III}$$

$$\text{Shaft III} \rightarrow \begin{cases} \longrightarrow \text{Dough feed screw 8} \\ \frac{\text{Gear wheel 6}}{\text{Gear wheel 7}} \rightarrow \text{Shaft IV} \end{cases}$$

$$\text{Shaft IV} \rightarrow \begin{cases} \longrightarrow \text{Dough plucking blade 9} \\ \frac{\text{Gear wheel 13}}{\text{Gear wheel 14}} \rightarrow \text{Shaft V} \end{cases}$$

$$\text{Shaft V} \rightarrow \begin{cases} \longrightarrow \text{Upper shaping roll 16} \\ \frac{\text{Gear wheel 17}}{\text{Gear wheel 18}} \rightarrow \frac{\text{Gear wheel 18}}{\text{Gear wheel 19}} \rightarrow \text{Shaft VIII} \rightarrow \begin{array}{c}\text{Lower}\\\text{shaping}\\\text{roll 20}\end{array} \\ \\ \frac{\text{Rope pulley 11}}{\text{Rope pulley 12}} \rightarrow \text{Shaft VI} \rightarrow \text{Dusting flour brush 15} \end{cases}$$

Scheme 3.1.   Transmission scheme for steamed bun maker.

3.5 OPERATION

To feed the steamed bun maker, the dough must be well mixed and cut into cubes of 1.0 to 1.5 kg. It is then added intermittently and well-distributed into the dough hopper where the dough level should be kept above the horizontal axis of the dough plucking blade to maintain a stable pressure and to ensure that there is enough dough for the blade to pluck to the screw chamber so that a consistent flow of dough is achieved so that it is possible to adjust the weight of each dough piece precisely. The adjustment can be carried out in two ways. Firstly, it is obtained by turning the handwheel (22) (Fig. 3.4) to adjust the gap between the movable valve and the screw sleeve, turning clockwise to decrease the dough piece weight, anticlockwise to increase the weight. Secondly, turning the handwheel (5) (Fig. 3.4) to change the pushing pressure to the dough by the dough feed screw, which results from the change of the length of the compression (working) section, turning clockwise to increase the weight, anticlockwise to decrease the weight.

The relative position of the spiral moulding slots of the two rolls is also very important for the appearance of the dough balls. If the ball has got a tail, open the right cover of the side wall and loosen the lock nut, pull the idler gear (18) out (Fig. 3.4) to allow the lower shaping roll to be turned clockwise by one or two or three teeth, or anticlockwise, then re-engage the idler gear before starting the machine to test the dough ball shaping. Generally speaking, the maximum turning is three teeth, and once the desired relative position of the moulding slots of the two rolls is achieved, it is not necessary to make any adjustment later.

The well-rounded dough pieces should rest for about 15 to 40 minutes before steaming. The resting time is dependent on the softness of the dough; the harder the dough is, the longer the rest needed.

Like any other food machine, hygiene must be maintained at all times. After each shift, the dough outlet on the screw sleeve should be dismantled for cleaning, and the shaping rolls, the dough hopper, dough screw, and the screw sleeve also should be cleaned to prevent any residual dough from drying and blocking the outlet and affecting the quality of the product.

Before operation the lubrication oil should be topped up in the three sets of transmission gears. The gears driving the two shaping rolls must be lubricated only with edible oil, to avoid food pollution. The bearings can be oiled once a quarter, with calcium base lubricant.

4

Dumpling manufacture

4.1 INTRODUCTION

A 'dumpling' is called 'Jiao Zi' in Chinese. It is a favourite food in China. Traditionally, 'Dumpling' is a symbol of reunion, happiness, and safety. When one of your relatives or friends is going on a journey, you may make dumplings as a send-off dinner, which expresses your wish that the traveller will arrive 'safe and sound'. The Chinese New Year is the most important festival in China, just like Christmas in many Western Countries. This is a day of family reunion, at its midnight advent the whole family always sit together, around a big table, and have their New Year dumplings after setting off fireworks and firecrackers. Here, the 'dumpling' symbolizes reunion, harmony, and happiness. As well as the specialized meaning at a special time, people often make dumplings when receiving family guests or for their own enjoyment, since it is such a delicious food.

About ten years ago, more than ninety per cent of dumplings were made by hand, and because of the time-consuming procedure people did not have them very often. Along with the accomplishment of dumpling-making mechanization, people can now buy frozen dumplings in the supermarkets. Dumplings more and more often appear on the domestic dining tables. Many restaurants make dumplings as a snack (quick meal) which is becoming more and more popular.

A Chinese dumpling comprises two parts; one is the casing dough, the other is the filling, a complex mixture of minced meat or seafood, vegetables, salt, edible oil, ginger, pepper powder, and other flavourings. The dough is a simple mixture of flour and water at a ratio of 100 (flour):38–40 (water).

For hand-made dumplings, the casing material (dough) should be rolled into small round sheets with a weight of about ten grams. For example, 100 g of flour is mixed with 40 g of water, the dough will be 140 g which can be divided into 12 to 18 pieces. Figure (4.1) shows the manufacturing procedure.

For a typical dumpling production line, a vegetable-washer, processor (cutter), dehydrator, and two mixers, of which one is for dough-mixing and the other for stuffing-mixting, are needed. A production rate of 7200 dumplings per hour can be achieved. Fig. 4.1 shows the process compared with dumpling-making by hand.

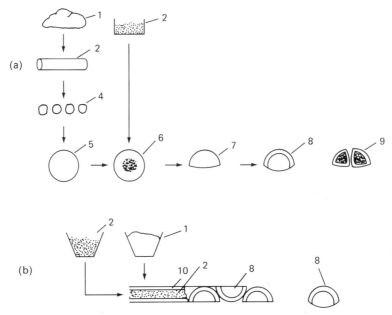

Fig. 4.1   Dumpling making processes (a) by hand, (b) by machine (1) Dough, (2) Stuffing, (3) Dough roll, (4) Dough pieces, (5) Circular dough sheet, (6) Filling, (7) Fold-over, (8) Dumpling, (9) Section through dumpling, (10) Dough pipe.

The first dumpling maker was invented 33 years ago. It was larger and more complicated than present-day machines. Its products were inferior to those made by hand. After more than thirty years of unremitting effort, unceasing technological innovation, and steady improvement, the fifth generation has been developed. Fig. 4.2 shows a typical dumpling maker, and Table 4.1 gives specifications of this kind of machine.

4.2   CONSTRUCTION

This automatic machine consists of a main body mounted on castors which allow the machine to be moved easily to any suitable place for either production or during cleaning. Mounted on the main body are the dough feed and stuffing feed assemblies as well as the rolling-and-cutting assembly for shaping. The machine is driven by a motor through belt and pulley, worm gear, vibrator mechanism, and a series of gear trains.

4.2.1   Dough feed assembly

Fig. 4.3 illustrates this system, which consists of a dough tray (1), dough hopper (2) and chute (14) with a dough feed screw (5) in it, inner nozzle (10) and outer nozzle (9) and locking and adjusting nuts (8), (6), (13).

The dough feed screw is a single thread hollow conveyor and a conical front end tapered 1/10 for gradually reducing the working space of the spiral grooves (slots) and

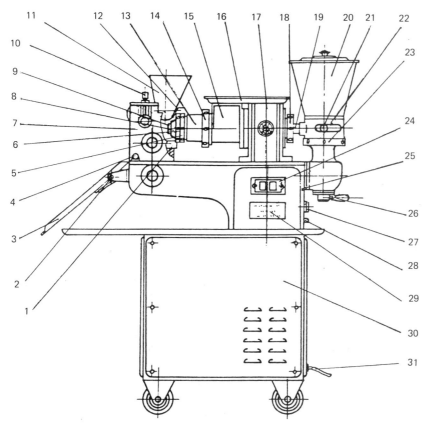

Fig. 4.2   A modern dumpling maker (Courtesy of Harbin Commercial Machinery General Factory, China) (1) Flour tray, (2) Vibration lever, (3) Chute, (4) Locating pin of forming frame, (5), (6) Nuts, (7) Forming frame, (8) Outer nozzle, (9) Inner nozzle, (10) Carriage bolt, (11) Flour hopper, (12) Dusting flour adjusting plate, (13) Dough screw sleeve, (14) Locking nut, (15) Dough hopper 2nd chamber, (16) Dough tray, (17) Gear box, (18) Locking nut, (19) Stuffing (filling) pipe, (20) Stuffing hopper, (21) Adjustment handle, (22) Vane (paddle) pump, (23) Locking nut, (24) Control panel, (25) Cast body, (26) Clutch handle, (27) Lubrication indicator, (28) Oil draining screw, (29) Digital display, (30) Machine frame, with motor inside, (31) Electric power inlet.

progressively increasing the pressure transmitted to the dough to meet the technological need of the later formation of a continuous and even dough tube. The inside hollow cylinder of the dough feed screw allows the stuffing pipe to be set into it, by which the stuffing will be filled into the dough tube as it is extruded through a compound outlet of the inner and outer nozzle.

There are two types of inner nozzle, as shown in Figs 4.3 and 4.9.

The first (Fig. 4.3) is fixed directly on the stuffing pipe by two locating screws. In this way the dough, conveyed by the dough feed screw, passes through the ringlike floating support (7) and thence into the space between the inner and outer nozzles (10), (9)

Table 4.1   Specifications of dumpling makers (Courtesy of Harbin Food Machinery Factory and Harbin Commercial Machinery General Factory, China)

No.	1	2	3
Capacity (pieces/h)	7200	6000	3600
Scaling range (g/piece)	13–20	12–18	13–20
Dumpling standard (dumplings/kg flour)	140–160	Max. 160	Max. 160
Motor	1.1 kW, 380 V 3 phase	750 W, 220 V	550 W, 220 V/380 V 50 Hz/60 Hz
Dimensions (mm)	990 × 470 × 1150	640 × 350 × 600	550 × 240 × 550
Net weight (kg)	160	80	52.5

whence the dough is pushed by a follower and extruded out of the narrow annular gap. By this time, a thin smooth dough tube is formed.

The second type (Figure 4.9) of inner nozzle is conical at one end, and at the other end is a ring with six kidney-like holes. The holes are arranged to coincide in two cycles so that the dough can pass through. The conical end cooperates with the outer nozzle to form a conical gap for the dough to pass through and to be shaped into a tube. In this way, the dough is first cut into six strips interlocked into each other in two

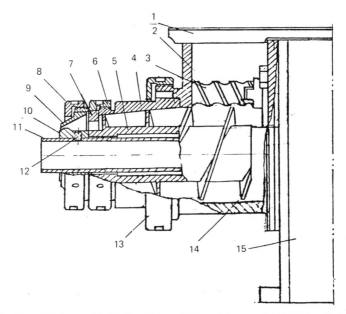

Fig. 4.3   Dough feed assembly (1) Dough tray, (2) Dough hopper, (3) Stability roll, (4) Screw sleeve, (5) Dough feed screw, (6) Locking nut, (7) Ring, (8) Locking nut, (9) Outer nozzle, (10) Inner nozzle, (11) Stuffing pipe, (12) Screw nail, (13) Locking nut, (14) Dough chute, (15) Gearbox.

layers as they pass through the six holes on the large end of the inner nozzle which is pressed against the dough screw sleeve (6) by nut (5) (Figure 4.9) without any other location mechanisms; Second, the six interlocked strips pushed by their followers move further and pass through the conical gap between the inner and outer nozzle (4), (2) where the separate strips are extruded into a thin seamless dough tube.

Comparing these two types of tube-forming assembly, the former type has some advantages since its passage is clearer, so that the dough is extruded at a lower temperature than in the latter type, whose passage is divided into six holes by the larger end of the inner nozzle so that the friction is increased when the dough passes through from the dough chamber to the nozzle unit, thus raising the temperature of both the dough and the machine parts. Furthermore, there is a potential danger, in the latter type, that the dough will have some hidden cracks which could appear in the next stage of dumping moulding. Too high a temperature is also harmful to the quality of the final product, especially the taste.

On the dough chamber is mounted a dough hopper which is equipped with a dough tray at the top and a stability roll which rotates in the opposite direction from the dough screw, to prevent the dough from bridging at the hopper entrance.

4.2.2 Stuffing feed assembly

Usually, there are two types of stuffing feed unit; the first is a combination of a screw feeder and a gear pump, and the second is composed of a screw feeder and a vane (paddle) pump. Experience has made it clear that the vane pump is better for keeping the original colour, flavour, and taste of the stuffing than the gear pump. Most dumpling makers are therefore equipped with the second combination. Figure 4.4 shows the form and working principle.

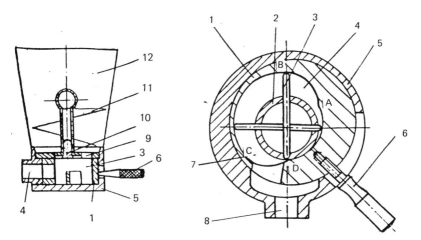

Fig. 4.4   Structure and principle of stuffing feeder (1) Stator, (2) Rotor, (3) Vane (paddle, wing), (4) Suction chamber, (5) Pump body, (6) Adjusting handle, (7) Compression chamber, (8) Outlet, (9) Pump inlet, (10) Shaft, (11) Stuffing feed screw, (12) Stuffing hopper.

The vane pump is a displacement pump noted for high pressure, stable flow rate, and precise rationing. It is therefore widely used in the food industry for delivering food materials. It consists of a rotor (2) on which are fixed the vanes (3) (paddle or wing) movable in two slots, a stator (1), and an adjusting handle (6). The stuffing hopper (12) is fixed to the pump body (5). The stuffing feed screw (10) is mounted on the same shaft (10) as the rotor, so that they rotate at the same speed. The screw forces the stuffing material into the pump and keeps a stable feeding rate which makes up for any deficiency due to insufficient filling that results from undesirable fluidity of the loose materials, or inadequate pump force, or other difficulties.

As shown in Figure 4.4, the inner wall of the stator (1) comprises of four zones, AB, BC, CD, and DA, with different curves. A pair of wings (3) set in the slots of rotor (2) divide the space between the inner side wall of the stator and the outer side wall of the rotor into four chambers. AB is suction chamber where the stuffing is sucked in through the entrance. CD is the compression chamber where the stuffing is compressed, and BC and DA are defined chambers where the cells are closed and neither entering nor expelling of the stuffing takes place. When the vertical feed screw rotates, the stuffing material is conveyed to the inlet of the pump and sucked into the suction chamber AB. The rotor coaxial with the feed screw causes the wings (vanes) (3) to rotate as well as to slide in their slots, forced by the side wall of the stator (1). Along with the movement of the wings, the suction chamber progressively increases in volume. When its maximum volume is achieved, the wings rotate and carry the stuffing through the BC section into the compression chamber (7) where the stuffing is compressed and forced through the outlet (8) into the horizontal filling pipe which is inside the dough feed screw. The continuous rotation of the screw feeder and the pump wings on the rotor make the filling material move continuously in and out of the pump, into the filling pipe which produces a dough tube filled with stuffing.

4.2.3   Moulding assembly

This assembly (Fig. 4.5) is composed of a lower roll (5) with a scraper (6), and a moulding roll (4) with a scraper (3), above which is the flour dusting device. A flour

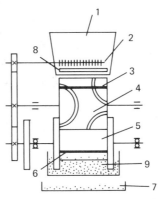

Fig. 4.5   Moulding assembly (1) Flour hopper, (2) Flour brush, (3) Moulding roll scraper, (4) Moulding roll, (5) Lower roll, (6) Lower roll scraper, (7) Excess flour collector, (8) Dusting flow adjusting plate, (9) Bottom tray.

tray (7) is placed below the lower roll (5) to collect the excess flour dusted from the flour hopper (1) by the rotary flour brush (2). Under the lower roll (5) is a dusting flour tray (9) full of dried flour which allows the roll always to be coated with a thin flour film to prevent any sticking as its surface passes through the flour mass.

The lower roll is a smooth cylinder case-hardened to improve its wearability. It is coated with Teflon to prevent sticking. The moulding roll is a hollow cylinder with dumpling-shaped engraved cutting edges which are staggered on the cylindrical surface. The semicircular areas, where the filled part of the dumpling will be, is hollowed out to allow the filled tube to have enough free space to prevent unnecessary contact as cutting is carried out.

4.3 SHAPING PRINCIPLE

As shown in Fig. 4.6, the filled tube (17) is continuously extruded from the inner and outer nozzle unit, moves forward, and comes into contact with the moulding roll (9) and lower roll (10), while the rotary brush (5) dusts flour on its surface to prevent sticking of the dough, which, as a mixture of flour and water, always tends to stick on a metal surface. As the filled tube is engaged, moulding begins.

The dumpling moulding process has two parts, printing and cutting, carried out at the same time. The filled tube is printed into dumplings by the engraving which makes the inside stuffing move to the centre part of the later dumplings so that the semi-circular ring is cleared and pressed together without any stuffing leakage. The

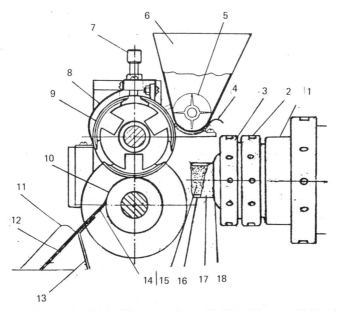

Fig. 4.6   Moulding device (1) Moulding screw sleeve, (2), (3) Locking nuts, (4) Dusting flour adjusting plate, (5) Brush, (6) Flour hopper, (7) Carriage bolt, (8) Moulding roll scraper, (9) Moulding roll, (10) Lower roll, (11) Chute, (12) Sieve, (13) Vibration lever, (14) Scraper, (15) Stuffing, (16) Dough tube, (17) Filled tube, (18) Outer nozzle.

dumpling is cut by the engraver's sharp edge, which is tangential to the lower roll. As stated in section 4.2.3, the moulding roll is a hollow cylinder, with engravings shaped into a shallow semicircular ring, making the outline of the dumpling touch the filled tube. The dumpling-shaped hollows of the moulding roll allow the filled part of the dumpling to be free from any unnecessary pressure on it, which would be harmful to the products, while the other part is printed and cut.

It is necessary to mention that the scrapers are only for cleaning the upper and lower rolls. There are no reusable scraps, as there are in biscuit production. Since the dumpling dough is highly elastic and the tube is very thin, the filled tube is moulded into a staggered arrangement without any dough pieces left. Any fragments stuck on either the lower roll or the cutting edges of the moulding roll are cleared by the respective scraper and collected into the tray under the moulding assembly.

The newly moulded dumplings are received by the chute (11) which is fitted with a bottom sieve and vibrated by the lever (13) for shaking off the dusting flour for reuse. Too much flour on the products will thicken the boiling water as the dumplings are cooked, which is unfavourable, and which increases flour consumption. The scraper (14) is fixed in front of the sieve without any other support. Not only is it for cleaning the roll, but it acts as a bridge for dumplings to pass over.

4.4   TRANSMISSION SYSTEM

Fig. 4.7 is a schematic diagram of a typical dumpling maker. The flow sheet is shown in Scheme 4.1.

From motor (1), the rotation is transmitted through belt pulleys (2) and (3), worm gears (4) and (5), to shaft II from which the motion is further transmitted in three routes to shafts III, IV, and VI.

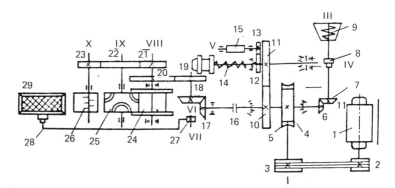

Fig. 4.7   Transmission system of a typical dumpling maker (Courtesy of Harbin Commercial Machinery General Factory, China) (1) Motor, (2), (3) Pulleys, (4) Worm, (5) Worm-gear, (6), (7) Bevel gears, (8) Stuffing pump, (9) Stuffing feed screw, (10), (11), (12), (13) Gear wheels, (14) Dough feed screw, (15) Stability roll, (16) Coupling, (17), (18) Bevel gears, (19), (20), (21), (22), (23) Gear wheels, (24) Lower roll, (25) Moulding roll, (26) Flour dusting brush, (27) Ratchet wheel, (28) Vibration lever, (29) Chute with sieve. (I)–(X) Shafts.

Motor 1 → $\dfrac{\text{Belt wheel 2}}{\text{Belt wheel 3}}$ → Shaft I → $\dfrac{\text{Worm 4}}{\text{Worm gear 5}}$ → Shaft II

Shaft II $\Bigg\{$

$\dfrac{\text{Bevel gear 6}}{\text{Bevel gear 7}}$ → Shaft III → Vertical screw 9 and vane pump 8

(to feed the stuffing)

$\dfrac{\text{Gear wheel 10}}{\text{Gear wheel 11}}$ → Shaft IV $\Bigg\{$

Dough feed screw (form dough tube)

$\dfrac{\text{Gear wheel 12}}{\text{Gear wheel 13}}$ → Shaft V → Stability roll 15

Coupling 16 → Shaft VI → $\dfrac{\text{Bevel gear 17}}{\text{Bevel gear 18}}$ → Shaft VII

Shaft VII $\Bigg\{$

$\dfrac{\text{Gear wheel 19}}{\text{Gear wheel 20}}$ → Shaft VIII $\Bigg\{$

→ lower roll 24

$\dfrac{\text{Gear wheel 21}}{\text{Gear wheel 22}}$ → Shaft IX

Ratchet wheel 27 → Vibration lever 28 → Chute 29 (to shake the dusted flour down)

Shaft IX $\Bigg\{$

→ Moulding roll 25 (to print and cut dumplings out of the filled tube)

$\dfrac{\text{Gear wheel 22}}{\text{Gear wheel 23}}$ → Shaft X → Dusting flour brush 26 (to distribute flour on the filled tube surface)

Scheme 4.1.   Transmission scheme for dumpling maker.

By means of the bevel gear wheels (6) and (7), the rotation direction is changed for the stuffing screw driven by shaft III to convey the stuffing vertically to the vane pump which supplies the stuffing to the filling tube at a preset rate through its rotary wings also driven by shaft III.

From shaft II, the rotation is transmitted through gear wheels (10) and (11) to shaft IV by which the dough feed screw (14) is actuated to convey the dough to the nozzles for the dough tube formation. At the same time the stability roll (15) is driven by shaft V through gear wheels (12) and (13) to press the dough into the dough chamber at a stable rate.

Through coupling (16), shaft VI, and another pair of bevel gears (17) and (18), the rotation is transmitted to shaft VII which delivers the motion in two branch routes:

one is through gear wheel (19) and (20) to shaft VIII; the other is through ratchet wheel (27) and vibration lever (28) to the chute (29) which is vibrated so that the flour dusted from the dumplings may be shaken down for reuse.

From shaft VIII, the rotation is further transmitted in two sub-branches: one is directly to the lower roll; the other is through gear wheels (21) and (22) to shaft IX which drives the moulding roll to print and cut out the dumplings from the filled tube on the lower roll, while the engaged gear wheels (22) and (23) actuate shaft X to make the brush (26) rotate and distribute flour on the filled tube surface to prevent sticking.

4.5   OPERATION

To operate a dumpling maker is not as easy as to run a mixer or a beater. For a new machine, running-in tests are necessary, and the dough tube thickness, the stuffing feed rate, and the dumpling size must be tested and preset before starting normal production. The procedure is as follows.

4.5.1   Starting the machine
Connect the mains to start the machine, running without any load. Attention should be paid to the rotating direction of the dough feed screw, which should be the same as that indicated on the dough hopper, otherwise it will not work properly, and the clutch handle (26) (Figure 4.2) should be set to the position 'OFF' before idling the machine.

During machine idling, the dough feed screw should be pushed to the end to avoid any contact and possible friction between the feed screw and the sleeve, otherwise the parts would wear.

4.5.2   Stuffing test (Fig. 4.2)
Before testing the stuffing feeding, it is necessary to pull out the locating pin (4) (Fig. 4.2) first, then turn the forming frame (7) aside clockwise 90°.

After filling the hopper (20) with stuffing, turn the clutch handle (26) to the position 'ON', set the adjustment handle (21) and fix it to test the stuffing flow. When the flow is constant, stable, without any interruption, and at the desired flow rate, keep it running for one or two minutes more, then turn handle (26) to the position 'OFF' to stop the feeding and wait for the later operation.

If the flow of stuffing becomes less , is interrupted, or even stops, turn off the machine immediately, examine the stuffing feed screw to check if the vane pump is blocked by some foreign body in the pump, or if the screw is in contact with fibrous material. Open the gear box for the stuffing feed to check whether the safety bolt of the bevel gear in the cast box (25) has sheared.

4.5.3   Dough tube test
Cut the well mixed dough into narrow bars, and put them into the dough hopper (15), (Fig. 4.2). Start the machine and check the dough tube. If the tube is not straight and smooth but crooked, loosen nuts (5) and (6) with a hook spanner and disassemble the inner and outer nozzles (9) and (8) to check if all parts are properly mounted. The

thickness of the dough tube can be regulated by adjusting nut (5). Turn it clockwise to get a thinner tube, anticlockwise to get a thicker tube.

4.5.4   Adjustment of dumpling size

After filling the upper hopper (11) (Fig. 4.2) and bottom tray (1) with flour, turn the forming frame (7) to its working position and fasten it by pushing in the locating pin (4). Start the machine to check the size and weight of the empty dumplings. Turn nut (18) clockwise to get larger dumplings, anticlockwise to get smaller ones.

If the dumpling edges are joined together, tighten up smoothly the two carriage bolts (10) to keep them separate. This can be done only very slightly and only while the machine is running.

4.5.5   Dumpling-making process

After the achievement of the desired dough pipe diameter, the proper flow rate of the stuffing, and the correct dumpling size, set the clutch handle (26) (Fig. 4.2) at position 'ON' so that the machine works continuously. Be sure to refill the hoppers and tray with dough, stuffing, and flour throughout working hours.

If some dumplings are broken, stop the machine immediately. This fault may result from an uncleaned mixer bowl in which the fresh dough is mixed with dried dough residue or other material which can block the nozzle and reduce the dough tube diameter when it is extruded from the nozzle unit. In this case, loosen the nut (5) a little to produce a thicker dough tube which allows the unwanted materials to be carried out; or disassemble the nozzles to get the sundries out. After the obstacles are removed, nut (5) should be returned to its original position. Then restart the machine.

To prevent the dumplings from sticking on the surface of moulding and lower rolls, the dough must not be soft. In general the water content is about 28%, and sufficient dried flour is needed as a release agent. If sticking takes place, the flour hopper should be checked to see if plate (12) is too far in, or the screen holes in the hopper bottom are blocked, or the flour dusting brush rotates abnormally, or there is insufficient flour and badly distributed. In any case, the trouble must be corrected.

4.5.6   Cleaning the machine

Hygiene is very important in food production. Since the dumpling is made from flour, meat, vegetables, edible oil, etc., which are all easily fermented, subject to deterioration, putrefaction, and even with the formation of acid or other toxic materials, thorough cleaning of the machine is especially necessary since even very little residue could corrode the machine parts and contaminate food. Hence, after each shift all parts in contact with the food materials (dough or stuffing) should be disassembled and cleaned thoroughly. The procedure is as follows.

(1) Cleaning of the dough feeder
Switch off to stop the machine, then pull out the locating pin (4) (Fig. 4.2) and turn the forming frame (7) clockwise 90°. Set the handle (26) to 'OFF'.

The following procedure is shown in Figure 4.8. Take off nut (1) and start the machine again. The outer nozzle (2) is pushed off by the extruded dough. Then dismount nut (3) to permit the inner nozzle (4) to be pulled out. Remove nut (5) to allow the screw sleeve (6) to be pushed out by the moving dough which is forced forward by the feed screw (7). Finally, stop the machine and pull dough screw (7) out for cleaning.

You may find that some parts are disassembled as the machine is running, while for others, the machine should be stopped for dismantling. This results from the dough of dumplings being just a mixture of flour and water without any oil or fat, and it is more adhesive and cohesive than biscuit and cake doughs. Furthermore, the feeding process is carried out under some pressure, which is caused by the feed screw and nozzles. The conical section of the screw makes the feeding pressure increase gradually, and the reduced space between the inner and outer nozzle (conical gap) also causes the dough to be extruded under pressure. Therefore these parts can have dough adhering tightly which makes the disassembly not as easy as in biscuit and cake makers. Hence, to start the dumpling maker and let the feed screw push the nozzle and sleeve out by the moving dough is a labour-saving way which is used in all dumpling makers. Since the screw works at a very low speed, there is no danger to the operator.

(2) Cleaning the dough hopper
As shown in Fig. 4.8, to clean the dough hopper, the screw (11) should be dismounted to allow the stability roll (8) to be removed from the hopper (9). Carefully clean the dough hopper and chamber and the stability roll by means of a plastic or hardwood scraper to avoid leaving any dried residue which will result in broken dumplings later. Cleaning must be carried out as soon as possible after stopping the machine, otherwise it will be hard to remove the stability roll out of the hopper since the dough

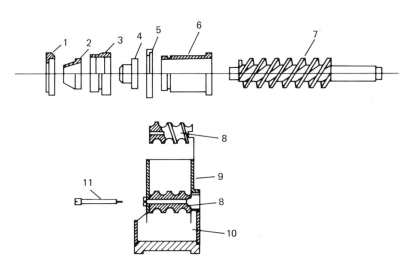

Fig. 4.8   Disassembly of the doughfeeder for cleaning and maintenance (Courtesy of Harbin Commercial Machinery General Factory, China) (1) Nut, (2) Outer nozzle, (3) Nut, (4) Inner nozzle, (5) Nut, (6) Screw sleeve, (7) Dough feed screw, (8) Stability roll, (9) Dough hopper, (10) Dough chamber, (11) Screw.

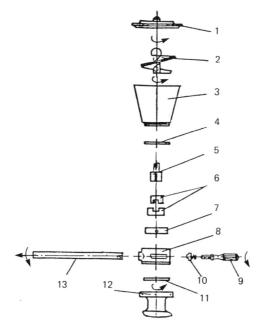

Fig. 4.9   Disassembling of the stuffing feeder for cleaning and maintenance (Courtesy of Harbin Commercial Machinery General Factory, China) (1) Lid of stuffing hopper, (2) Stuffing feed screw, (3) Stuffing hopper, (4) Upper movable plate, (5) Rotor, (6) Wing (vane), (7) Stator, (8) Stuffing pump body, (9) Adjusting handle, (10) Pillow, (11) Bottom Plate, (12) Nut, (13) Filling pipe.

is low in water content and will dry quickly and stick on the roll and the hopper, and join them together so that cleaning becomes less easy.

(3) Cleaning the stuffing feed assembly (Fig. 4.9)
Even though this assembly consists of many parts, it is quite easy to disassemble. Unlike the dough feeder, of which some parts should be pushed out by the moving dough with the machine running, to dismount the stuffing feeder simply follow Fig. 4.9. All parts can be easily taken out for cleaning. However, some suitable degreaser should be employed since the stuffing materials are rich in oil or fat.

The forming assembly does not need to be disassembled for cleaning. A brush will easily clean the moulding and bottom rolls and collect the flour into a tray for reuse, since the scrapers of the two rolls keep the moulding surfaces clean at all times.

5

Hun Tun manufacture

5.1 INTRODUCTION

Hun Tun is another stuffed snack, very popular in China. The differences between dumplings and Hun Tun are not only in the shape but also in the method of cooking, the hardness of the casing dough, and the consistency of the stuffing materials. Dumplings are cooked in boiling water or fried in a pan, but Hun Tun are usually cooked in a soup which may be chicken soup, or seafood soup, or some other tasty clear soup, dependent on personal taste. Dumplings are often eaten with a sauce (a mixture of vinegar and other flavourings such as mustard), while Hun Tun are already flavoured by the soup in which they have been cooked. A bowl of Hun Tun is always a good choice for breakfast or a midnight snack.

The casing material of Hun Tun is also a mixture of flour and water with a ratio of 100 (flour):35 or less (water), while for dumpling dough it is about 100:38-40, and for steamed buns, the dough is about 100:50. So the Hun Tun dough is the hardest, and no dusting flour is needed in Hun Tun production.

Usually, the stuffing for Hun Tun is a mixture of meat of seafood with various flavourings, but seldom vegetable, so that it is rather sticky.

Traditionally, the formation of Hun Tun is done by hand in the following stages:

(a)　prepare casing and filling materials—dough and stuffing;
(b)　roll the dough into a sheet with a thickness less than 2 mm;
(c)　cut the dough sheet into trapezium pieces with dimensions of about 60 × 60 × 110 mm (upper × height × bottom);
(d)　fill the piece of dough sheet with stuffing;
(e)　fold and close it.

The first Hun Tun-making machine, Hun Tun maker for short, was invented in about 1980. After several years of work by many technicians and engineers the newly developed Hun Tun maker is an advanced and ingeniously constructed machine in which the shaping process is an imitation of the manual method. It is about the size of a domestic microwave oven, but is very efficient. It is capable of forming 4920 pieces of Hun Tun per hour. The weight of the products can be changed easily within a range of 7 g to 15 g. This space-saving and energy-saving machine can be used in restaurants to

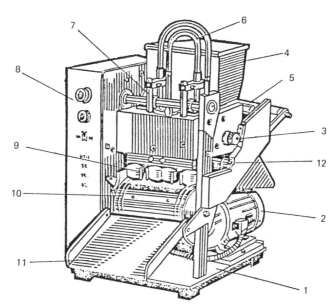

Fig. 5.1   Hun Tun maker (Courtesy of Wu Xi Supply and Marketing Factory, China) (1) Stand, (2) Motor, (3) Stuffing adjustment, (4) Stuffing hopper, (5) Dough sheet feed-in, (6) Compressed air supply, (7) Finishing punch, (8) Control panel, (9) Floating shaping rod, (10) Shaping roll, (11) Chute, (12) Finishing punch adjusting lever.

supply snacks, or in food factories to produce frozen Hun Tun for supermarkets. The electrical consumption of a maker is only 0.55 kW. Its appearance is shown in Fig. 5.1.

A Hun Tun production line is composed of a mincer, two mixers for mixing dough and stuffing respectively, a sheeter for supplying dough sheet to the Hun Tun maker, and a products receiving conveyor leading to panning for freezer or cooker.

5.2   CONSTRUCTION

The Hun Tun maker is rather small but fairly complicated. It includes a dough sheet feed mechanism, stuffing feed system, forming mechanism, compressed air system, transmission and control systems, etc. The lower water content, very smooth surface, and dense structure of the dough sheet, make it unnecessary for the Hun Tun maker to be equipped with a dusting flour device.

5.2.1   Dough sheet feeder
The dough sheet, which comes from an associated sheeter, passes into the sheet feed system of the machine (Fig. 5.2). It consists of four pairs of rolls: sheet levelling (flattening), longitudinal cutting, cross-cutting, and accelerating, which supply the next stages with 80 × 90 mm dough sheet pieces. To help the tendency for dough sheet pieces to self-slide during feeding, the four pairs of rolls are arranged in an inclined plane, 30° to the horizontal and joined by means of guide plates.

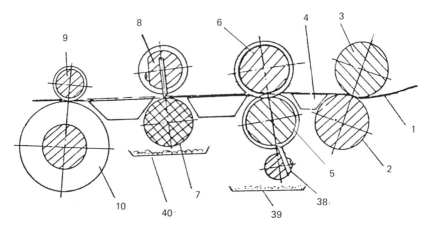

Fig. 5.2 Dough sheet feed system (1) Dough sheet, (2) Lower levelling roll, (3) Upper levelling roll, (4) Guide, (5) Lower longitudinal cut roll, (6) Upper longitudinal roll, (7) Lower roll, (8) Cross-cut roll, (9) Floating press roll, (10) Accelerating roll, (38) Scraper, (39) Scrap tray, (40) Residue tray.

(1) The levelling section
The dough sheet is very thin (about 1 mm in thickness and 200 mm in width), therefore if it passed directly from the sheeter to the cutting section, it would wrinkle and spoil the shape of the end-product.

This is obviated by the levelling rolls (2) and (3). The lower roll is driven by gears. The upper roll (3) floats on the lower roll (2) and is actuated by friction as the dough sheet is engaged and flattened.

(2) The longitudinal cut section
In this section, the upper roll (6) is fitted with three disk blades which match three circumferential grooves on the lower roll (5), which are cleaned by scrapers (38). Scraps are collected in the tray (39). When the flattened dough sheet is engaged, the blades cut the sheet (200 mm in width) into two standard dough strips (90 mm for each width) while the small side leftovers of the sheet are collected in the tray (40) for reuse.

(3) The cross-cutting section
The clockwise upper roll (8) is fitted with five knives parallel to the shaft axis. To save the knife edges from damage, the lower roll (7) is made of non-toxic rubber. As the two dough strips from the longitudinal cut section come into contact with the pair of cross-cut rolls (7) and (8), they are cut into rectangular pieces (80 × 90 mm). One revolution of roll (7) and (8) corresponds to ten small dough pieces since there are five knives to cut the two strips.

(4) The accelerating section
The rectangular dough pieces are fed continuously from the cross-cutting section. They must be spaced from one another before they go to the shaping plate of the filling section. Spacing is achieved by the action of rolls (9) and (10). The diameter of roll (10)

is twice that of the earlier rolls, and its rate of revolution is six times greater. By the time that the dough piece is nearly cut off from the strip, its front has been engaged by rolls (10) and floating roll (9), where weight helps to increase the transmitted force to the cut dough piece. As soon as the dough piece is cut off by the cross-cut rolls, rolls (10) and (9) accelerate the piece to a speed twelve times as great as its former speed, hence, a leap forward takes place. The rectangular dough piece flies onto the shaping plate, and there is a gap between it and the following piece, which is moving at its original speed. This space allows the dough piece on the shaping plate to wait for filling and folding.

5.2.2 Stuffing feed system
Fig. 5.3 illustrates this system, which comprises a hopper (13) fitted with a feed screw and a scraper at its lower part, stuffing fill tube (14), feed punch (15), adjusting rack and knob (22) and (44), and the compressed air system.

(1) Feeding
The stuffing feed screw is a half left and half right-handed one for conveying the stuffing from both sides of the hopper to its central part which is in the face of the stuffing outlet (18). Driven by its shaft (41), the feed screw (16) rotates at 40 rev/min, which matches the speed of the cut piece feeding. The scraper (17), also fixed on shaft (41) by the filling head screw (42), pushes the stuffing from the outlet of the hopper to the inlet of the stuffing tube through the stainless liner (45). The empty space left by the stuffing is immediately made up by means of the feed screw from two sides as the screw has a half right and half left-handed thread. The thread vanes and the scraper work together to feed the stuffing tube (14) continuously. The feeding rate can be easily adjusted by turning the knob (44) which drives the rack (22) forward and

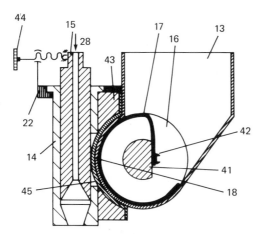

Fig. 5.3   Stuffing feed assembly (13) Stuffing hopper, (14) Stuffing tube, (15) Stuffing feed punch, (16) Left-hand, right-hand screw, (17) Scraper, (18) Stuffing outlet, (22) Stuffing adjusting rack, (28) Compressed air, (41) Shaft, (42) Filling head screw, (43) Connector, (44) Adjusting knob, (45) Stainless liner.

backward to make the engaged gear of the upper end of the stuffing tube revolve rightward or leftward so that the opening of the outlet (18) becomes larger or smaller.

(2) Separating
The cohesive stuffing in the stuffing tube (14) and in the hopper still coheres. To separate it into portions, the stuffing feed punch (15) is employed. When the punch, along with the connection body (43), moves downward and passes through the inlet of the tube, the stuffing is divided into two parts, of which one is pushed down to the bottom outlet of the tube and the other part is left in its original place.

(3) Blowing down
At this stage, compressed air is needed. The compressed air system is composed of an oscillate-type and single stage air pump, one-way valve, pipe, and control valve, etc.

Pushed by the stuffing feed punch (15), the stuffing arrives at the exit end of the tube (14). It is not able to fall down by gravity because of its adhesiveness. By then, the air cylinder is exactly in the predetermined position of its compression stroke. The normally closed valve in the air pipe is opened instantaneously to allow the compressed air from the cylinder to pass and go further into the small hole of the feed punch (15), and to blow the stuffing down onto the waiting cut dough piece on the shaping plate (12) (Fig. 5.4). The pressure of the compressed air is about 1.5 kg/cm^2, which is enough to blow all the stuffing off, leaving the exit end of the tube clear.

5.2.3   Forming mechanism
As shown in Fig. 5.4, the system mainly comprises a turn-up plate (11) driven by a cam-rack and gear mechanism, shaping roll (25), a finishing punch (quarter touching) (21), floating shaping rod (24), shaping guide (23), chute (27), step-by-step mechanism (48), and cam follower mechanism.

Through shaft (49), the shaping roll (25) is driven by the step-by-step mechanism (48). One revolution of the driving gear corresponds to a quarter of a circle rotation of the shaping roll. That is, every revolution of the driving gear makes the shaping roll revolve by 90° intermittently. The space between each motion allows the cut dough pieces on the shaping roll (25) to wait for the filling and folding process.

Two cams mounted on the shaft (49) inside the cylinder drive four pairs of followers with floating shaping rods (24) to move intermittently up and down for pushing the filled and folded pieces through the shaping plate aperture (12), and to be side-bended and finished by the shaping guide roll (23) and the finishing punch (21).

5.3   SHAPING PRINCIPLE

The shaping principle of Hun Tun is a rather complicated process which can be divided into five stages: location, first folding, second folding, side-bending, and finishing. To make it easier to understand, Fig. 5.5 shows the general process, while Fig. 5.6 shows the separate stages and the product shape at each stage. For clarity, a uniform numbering system is used in Figs 5.2, 5.3, 5.4, 5.5, and 5.6.

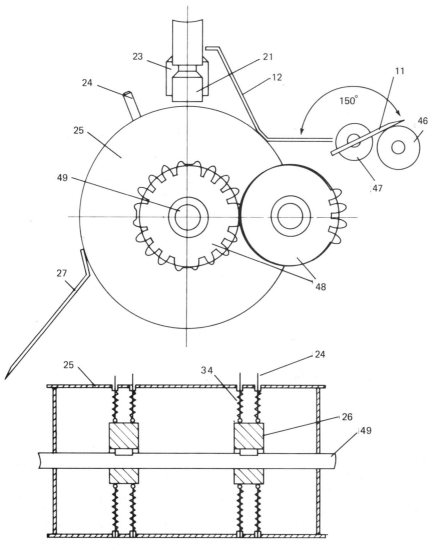

Fig. 5.4   Forming system (11) Turn-up plate, (12) Shaping plate, (21) Finishing punch, (23) Shaping guide, (24) Floating shaping push rod, (25) Shaping roll, (26) Cam, (27) Chute, (46) Locating roll, (47) Turn up plate shaft, (48) Step-by-step mechanism, (49) Shaping roll shaft.

5.3.1   Location

In Fig. 5.6a at the cut piece (10) delivered by the accelerating roll should be located on both the shaping plate (12) and the turn-up plate (11). To make the traditional Hun Tun, the shaping plate is 6–8 mm longer than the turn-up plate. The angled shape of (12) makes it possible to locate and shape the cut pieces.

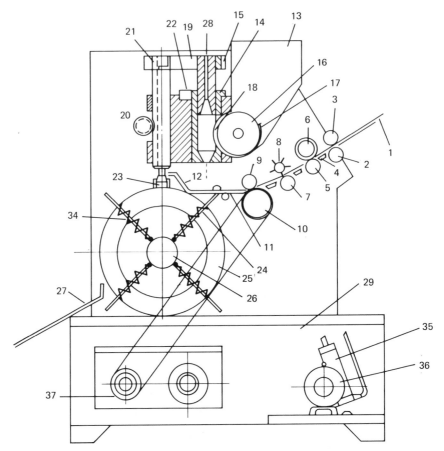

Fig. 5.5   Forming principle, general (Courtesy of Wu Xi Supply and Marketing Factory, China) (1) Dough piece, (2) Lower levelling roll, (3) Upper levelling roll (floating), (4) Guide, (5) Lower longitudinal cut roller, (6) Upper longitudinal cut roll, (7) Lower cross-cut roll, (8) Cross-cut roll, (9) Floating pressure roll, (10) Acceleration roll, (11) Turn-up plate, (12) Shaping plate, (13) Stuffing hopper, (14) Stuffing, (15) Stuffing feed punch, (16) Left-and-right hand screw, (17) Scraper, (18) Stuffing outlet, (19) Coupling plate, (20) Gear, (21) Finishing punch, (22) Stuffing adjustment rack, (23) Shaping guide, (24) Floating shaping push rod, (25) Shaping roll, (26) Cam, (27) Chute, (28) Compressed air inlet, (29) Machine body, (35) Air compressor, (36) Motor, (37) Transmission system.

5.3.2   First folding

As soon as the stuffing is blown down onto the dough piece, the turn-up plate (11) driven by cam (30) (Fig. 5.6b), rack (31), and gear (32), turns 150° upward so that the back end of the cut piece on plate (11) is folded over the leading part on the shaping plate (12), so that the stuffing is covered. To prevent the folded piece from re-opening, the cam (30) provides an idle stroke which causes plate (11) to rest for a preset time, thus maintaining pressure on the folded piece until its shape has stabilized. During the down stroke of rack (31), plate (11) turns back to its original position.

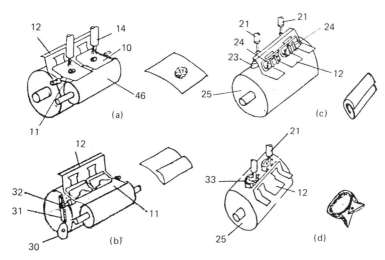

Fig. 5.6   The forming process (Courtesy of Wu Xi Supply and Marketing Machinery Factory, China) (10) Dough sheet, (11) Turn-up plate, (12), Shaping plate, (14) Stuffing tube, (21) Finishing punch, (23) Shaping guide, (24) Floating shaping push rod, (25) Shaping roll, (30) Cam, (31) Rack, (32) Gear, (33) Shaped Hun Tun, (46) Locating roll.

5.3.3   Second folding

After the plate (11) has returned, the shaping roll (25) (Fig. 5.6c), powered by step-by-step mechanism, revolves 90° intermittently, while the four pairs of shaping rods arranged equally around the circumference of roll (25) revolve at the same time. The revolving rods are also pushed radially out of the surface of roll (25). The projecting and revolving rods (24) make the simply-folded pieces move upward along the inclined plane of plate (12), which provides the second folding.

5.3.4   Side-bending (Fig. 5.6c and d)

The rods push the twice-folded piece further upwards (Fig. 5.6c,d). When the piece passes through the aperture in plate (12), its two sides are bent 90° into a U-shape under the guidance of the two guide shaping rolls (23). The distance between the two guide rolls is about the same as the width of a traditional Hun Tun. To prevent the dough piece from breaking in the bending process, the floating guide rolls revolve freely, so that only a gentle force is applied.

5.3.5   Sealing and corner forming

It now remains to seal the two ends of the dough envelope and shape its four corners, thus producing the traditional shape of Hun Tun.

After turning 90°, the shaping roll (25) stops, and the partly formed Hun Tun is pushed by the shaping rods (24) into position exactly below the finishing punch (21) (Fig. 5.6d). The punch, driven by a crank-slider mechanism, is rapidly applied to the U-shaped piece, pressing the edges of the open ends together and firmly bonding them. The four corners of the Hun Tun are shaped by the same operation.

The punch (21) has a spring buffer which prevents the Hun Tun piece being damaged by impact and keeps pressure on the overlaps to ensure their stability. When punch (21) returns to its initial position, the shaping roll (25) starts to turn a further 90°, and the shaping rods (24) also turn by 90°, moving in their down stroke along the surface of the rotating cam (26) (Fig. 5.5). The finished Hun Tun is delivered by the chute (27) to a web or pan for freezing or cooking.

5.4   TRANSMISSION

A typical Hun Tun maker is driven by a 0.55 kW electric motor through more than seventeen shafts, a series of belt and wheel, gear trains, cam-follower, rack and pinion, crank and connecting rod, and other systems. There are twelve transmission routes, branches, and sub-branches.

From motor (1) (Fig. 5.7), the rotation is transmitted to the spindle, shaft IV, through belt and wheel (2), (3), and (4), a mechanical reducer (5) and chain gearing (6), (7), and (8). The spindle IV transmits the power via four routes:

(A)   From gear train (9) and (10) the shaft IX receives rotary motion and transmits it via four branches:

(a)   By means of gear train (24), (25), (26), (33) and the intermediate shaft XIV, the levelling roll 32 actuates the floating roll 31 by friction to make the dough sheet from a matching sheeter flat for the next stage of the process.

(b)   Driven by the gear train (24), (25), (26), (29), and (30) through shaft XII, the roll (27) and (28) rotate in reverse directions to cut the engaged flattened dough sheet into two standard strips by the two disk cutters mounted on roll (27).

(c)   Roll (22) and (23) driven by gear train (50), (51), cross-cut the dough sheet strips into the required small rectangular pieces by means of the revolving blades axially mounted on shaft roll (22).

(d)   From gear (24), (41), and (42), the rotation is transmitted by shaft XVI to the stuffing feed screw (43), which conveys the stuffing to its middle part for the scraper (44) to push it further into the stuffing tube through the outlet (45) at the lower part of the hopper.

A crank gear (10) drives the finishing punch (37) via the rack (40) to reciprocate vertically to bring about the final formation of the Hun Tun by pressing its two ends together. Through pinion (39), shaft XI, pinion (34), and gear rack (35), the rack (40) also actuates the stuffing punch (38) to move reciprocally to push the stuffing out of the stuffing tube exit.

(B)   Along with the floating roll (21), roll (20) on shaft IV rotates at a higher speed than the other rolls to accelerate the cut dough pieces, delivering them to the shaping plate above the shaping roll (18) to wait for filling without interference from the pieces that follow it.

(C)   The one-way valve of the compressed air system is opened by a connecting rod driven by gear (9) and rank (11), which is also a gear assembly, to blow the preset

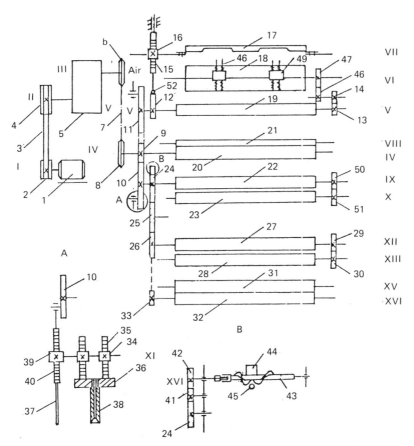

Fig. 5.7 Transmission system of Hun Tun maker (1) Motor, (2) Pulley, (3) Belt, (4) Pulley, (5) Reducer, (6) Chain gear, (7) Chain, (8) Chain gear, (9) Gear, (10) Gear (crank), (11) Gear (crank), (12) Cam, (13), (14), Gears, (15) Rack, (16) Pinion, (17) Turn-up plate, (18) Shaping roll, (19) Locating roll, (20) Accelerating roll, (21) Floating roll, (22) Cross-cut roll, (23) Lower roll, (24), (25), (26) Gears, (27) Longitudinal cut roll, (28) Longitudinal cut lower roll, (29), (30) Gears, (31) floating roll, (32) Levelling roll, (33) Gear, (34) Pinion, (35) Rack, (36) Connection piece, (37) Finishing punch, (38) Stuffing punch, (39) Pinion, (40) Rack, (41), (42) Gears, (43) Left- and right-hand screw, (44) Scraper, (45) Stuffing outlet, (46), (47) Gears, (48) Floating shaping rod, (49) Cam, (50), (51) Gears, (52) Cam follower. (I)–(XVI) Shafts.

stuffing off the stuffing tube exit. The stuffing drops down onto the waiting dough piece. When the crank arrives at its top position, the valve is closed until its next opening.

(D) From gear (9) and (10), rotation is transmitted in three ways via shaft V:

(a) Driven by cam (12) and circular follower (52) through rack (15) and pinion (16), the turn-up plate (17) on shaft VII swings reciprocally to turn the dough piece up to the preset position on the shaping plate and to maintain pressure on it to prevent the dough piece from re-opening. This movement is determined by the shape of the cam.

(b)   Through gears (13), (14), and the step-by-step gear mechanism (46) and (47), shaft VI is rotated, and the cams push the floating rods (cam followers) out in synchronization with the movement of the shaping roll whose step-by-step rotation is controlled by incomplete-toothed gear mechanism (46) and (47). This allows the dough pieces to be filled, folded, and final-punched step by step so that the Hun Tun receives its final form.

Roll (19) on shaft V is for locating and stabilizing the turn-up plate (17). The turning motion of the shaft can buffer the strike when the plate swings back against it. The transmission scheme is shown in Scheme 5.1.

$$\text{Motor 1} \rightarrow \frac{\text{Pulley 2}}{\text{Pulley 4}} \rightarrow \text{Reducer 5} \rightarrow \frac{\text{Chain gear 6}}{\text{Chain gear 8}} \rightarrow \text{Shaft IV}$$

Shaft IV $\Big\{$

$\dfrac{\text{Accelerator roll 20}}{\text{Floating roll 21}} \rightarrow$ Dough cut-piece to be accelerated to shaping plate above roll 18

$\dfrac{\text{Gear 9}}{\text{Crank 11 (gear)}} \rightarrow$ Compressed air system $\rightarrow$ Stuffing drops on dough piece

$\dfrac{\text{Gear 9}}{\text{Gear 11}}$ Shaft V $\rightarrow$ $\Big\{$

  Locating roll 19 $\rightarrow$ Turn-up plate 17 to be located

  $\dfrac{\text{Gear 13}}{\text{Gear 14}} \rightarrow \dfrac{\text{Gear 46}}{\text{Gear 47}} \rightarrow$ Shaft VI

  $\dfrac{\text{Cam 12}}{\text{Follower 52}} \rightarrow \dfrac{\text{Rack 15}}{\text{Pinion 16}} \rightarrow$ Shaft VII

$\dfrac{\text{Gear 9}}{\text{Gear 10}} \rightarrow$ $\Big\{$

  Shaft IX $\Big\{$

    $\dfrac{\text{Gear 24}}{\text{Gear 25}} \rightarrow \dfrac{\text{Gear 25}}{\text{Gear 26}} \rightarrow$ Shaft XII

    $\dfrac{\text{Gear 26}}{\text{Gear 33}} \rightarrow$ Shaft XIV

    $\dfrac{\text{Gear 50}}{\text{Gear 51}} \rightarrow$ Roll 23 $\Big\}$ Dough sheet crosscut

    $\rightarrow$ Roll 22

    $\dfrac{\text{Gear 24}}{\text{Gear 41}} \rightarrow \dfrac{\text{Gear 41}}{\text{Gear 42}} \rightarrow$ Shaft XVI

  Crank (gear) 10 $\Big\{$

    $\dfrac{\text{Rack 40}}{\text{Pinion 39}} \rightarrow$ Shaft XI

    Rack 40 $\rightarrow$ Finishing punch 37 (Hun Tun shaped)

Shaft VII → Turn-up plate 17 → Dough sheet to be turned up

Shaft VI → $\begin{cases} \text{Shaping roll 18 revolving step by step} \\ \\ \text{Cam 49} \rightarrow \text{Spring rod 48 to be pushed out and in} \end{cases}$ $\left.\begin{array}{l} \\ \\ \end{array}\right\}$ Dough piece to be folded

Shaft XII → $\begin{cases} \rightarrow \text{Roll 27} \\ \dfrac{\text{Gear 29}}{\text{Gear 30}} \rightarrow \text{Roll 28} \end{cases}$ Dough sheet to be longitudinally cut

Shaft XIV → Leveling roll 32 → Floating roll 31 → Dough sheet
to be flattened

Shaft XVI → Screw 43 → Stuffing feeding

Shaft XI → $\dfrac{\text{Pinion 34}}{\text{Rack 35}}$ → Stuffing punch 38 → Stuffing to be pushed out

Scheme 5.1.   Transmission scheme for Hun Tun maker.

5.5   OPERATION

The Hun Tun maker is a small and complex machine. However, it is very easy to operate. The procedure is as follows.

Firstly, the air compression system should be checked to see if the air pipe is clear and that the valve can be precisely controlled.

Secondly, the dough sheet feed system should be tested to see that the dough sheet runs smoothly and the pieces are cut to the required size and are in the right position. Usually, the finishing punch needs a regulation by means of lever (12) (Fig. 5.1) to adjust its stroke, since a short stroke will lead to incompletely shaped products, and a longer stroke will cause severe damage to the punch and to the shaping roll, holed products, and dough sticking on the shaping roll.

Thirdly, fill the hopper with enough stuffing at a sufficient level to ensure stable feeding, since the stuffing applies a force to the lower part of the hopper which affects the feeding rate. Under normal conditions, the amount of stuffing is adjusted by turning the knob (3) (Fig. 5.1). The stuffing feed assembly is the part most in need of cleaning. As the feed screw is engaged with driving shaft XVI through a bayonet catch (Fig. 5.7), it can be easily disassembled for cleaning with a non-toxic de-oiling solution, because the stuffing is usually rich in oil or fat and very viscous. Cleaning should be carried out as soon as possible after each shift to prevent any pollution by the readily decayed stuffing materials. The scraps collected in tray (40) (Fig. 5.2) can be taken to the dough mixer for re-use, while other residues in tray (39) (Fig. 5.2) are scanty and useless. However, both trays should be cleaned after each shift.

Part III

Bread-making machines

6

Dividers

6.1   INTRODUCTION

The function of a divider is to cut the fermented dough for bread into preset sized pieces. Dividing accuracy is a key factor in the economic, profitable operation of all bakeries. Because of the elastic and cohesive nature of the bread dough it has proved difficult to design an apparatus to subdivide the dough on a weight basis. Consequently, all dividers at in use at present operate on a volumetric basis [1].

The simplest divider was a group of framed cutters to divide a flattened dough into a group of the same preset sized dough pieces. Afterwards, they were rounded by hand. Inspired by this principle, a half-mechanical and half-manual machine was invented which combined the cutting and rounding. In this machine, a flattened thick dough piece was put onto the working table by hand. Then, a controlled cutting head was driven downward by a slider crank chain to cut the dough into pieces. The controlled knives formed many cells containing the divided pieces. As the table of the working chamber revolved, the pieces were rounded by means of centrifugal force, gravity, and friction. When a round shape and a compact structure were achieved, the cutting head returned to its original position to allow the rounded dough pieces to be taken out for the next process. No doubt, this was a great advance from manual labour to mechanization. This type of machine can still be found in some bakery shops and laboratories. But it is certainly unsatisfactory for today's bakery industry.

To meet the requirement of high efficiency, modern dividers have been developed which can cut the soft sticky fermented dough automatically and accurately. Fig. 6.1 illustrates a common type divider.

6.2   CONSTRUCTION

6.2.1   Hopper

The machine, which is movable, comprises a dough hopper, a dividing head, a special oil distribution system, and transmission, control, and power supply systems.

The hopper of early dividers was fixed by bolts on the top of the dividing assembly. But in modern machines, the hopper can hinge down to allow cleaning from floor

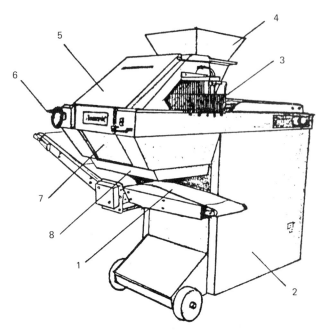

Fig. 6.1   A typical divider (Courtesy of APV Baker, UK) (1) Delivery conveyor, (2)
Transmission gear and power supply, (3) Lubricating device, (4) Dough hopper, (5) Guard,
(6) Adjusting handwheel, (7) Dividing head, (8) Dough piece outlet.

level. The hopper is coated internally with PTFE to ensure that the dough is fed
smoothly into the divider and to make cleaning of the hopper easier.

6.2.2   Dividing head
The dividing assembly (Fig. 6.2) is the heart of the machine. It contains a chamber
where the dough is sucked from the bottom of the hopper and compressed into the
division box by the reciprocating ram and knife. Functionally, the chamber is referred
to as the compression or suction chamber. Associated with it are a division box which
consists of pockets with dies in each, a discharge block, and a front plate fitted as a
cantilevered head which gives easy access to the division box, dies, and the receiving
(primary) conveyor for either maintenance or cleaning. To satisfy today's hygiene
standards, all parts in contact with the dough are either made of high quality stainless
steel or covered with sophisticated machined polymers, which are stable, highly
resistant to wear, and easy to clean. These qualities are particularly important for the
front plate, as its surfaces must run smoothly over each other, reducing the friction
loads to improve the operation conditions and prolong the working life of the
machine. A modern divider is always equipped with a spring loading system to
maintain the pressure on the front plate between the dividing box and the base of the
hopper, creating a good seal, and compensating for any wear.
    The highly viscous character of the bread dough makes it impossible to divide the
dough into pieces at the high speed used in biscuit-cutting. The limiting speed at

which a divider may be effectively operated is about 20 to 28 strokes per minute. Modern machines are equipped with 2 to 8 pockets, consequently, 2 to 8 pieces of dough are produced for each stroke of the ram. For example, a seven-pocket divider running at a speed of 20 strokes per minute will deliver 8400 dough pieces per hour. For machines of the same size, the more pockets that are provided, the smaller the dough pieces produced. Thus, a four-pocket machine has a size of 250 to 700 g, compared with only 40 to 200 g for a seven-pocket machine. Table 6.1 shows specifications of some large dividers.

In some machines the pockets are fitted in the division box, while in others they form part of the box. The volume of the pockets is adjusted by changing the die (piston) travel (depth) to allow variation in the weight of the finished loaf and to permit compensation for changes in dough density. The weight of the dough pieces becomes progressively lower from the start to the end of dividing, because of continuing fermentation accompanied by a steady generation of carbon dioxide gas in the dough. From this point of view, the dividing operation should be performed as rapidly as possible. However, it is essential that a variable speed drive be equipped to match the stepless adjustment of scaling rate and stroke speed, as well as the following rounder which needs a uniform feed by the primary conveyor receiving the divided dough pieces under the dividing head.

6.2.3 Lubrication system

Dough dividers have a special lubrication system which differs from that of the other food machinery discussed in this book. The freshly divided dough pieces are extremely sticky, and they tend to adhere to any surfaces with which they come into contact unless a suitable release agent is used. Oil, rather than flour, is the only choice. The oil can not only lubricate the mechanical parts of the machine, which come in contact

Table 6.1   Specifications of some large dividers (Courtesy of Tokyo Maruichi Shoji Co., Ltd, Japan)

Pocket number	2	4	5	6A	6B	7	7B
Scaling range (g)†	700– 1400	250– 700	200– 560	160– 380	180– 480	40– 200	120– 320
Capacity (pieces/h)	1200– 2400	2400– 4800	3000– 6000	3600– 7200	3600– 7200	4200– 8400	4200– 8400
Stroke			20 times/min (stepless)				
Hopper capacity			170 litre (standard)				
Motor			2.2 kW				
Dimensions			1170 × 1670 × 2010 (mm)				
Net weight			2600 kg				

†Scaling range is at the dough specific gravity of 1.0..

with the dough, but also form a seal between the dough box and the dividing head. From the food hygiene aspect, edible oil should be the best choice, but it has been found that vegetable oils leave a gummy film which may cause excessive friction or even bind the close tolerance moving parts of the machine [1]. Hence, a special grade of mineral oil has been used for many years. This divider oil is colourless, odourless, tasteless, and safe. But it can be used only at a level not exceeding 0.15%, according to the Food Additive Regulations, for it would be taken up by the dough and would remain in the finished loaves [1].

There are two types of oil system. In one system, the oil is fed by gravity from an elevated container through tubes to the critical surfaces, and the surplus oil is collected into a tray at a lower level. In another system the oil is pressure-fed by a pump to the dividing mechanism to ensure smooth running for long periods without stopping for cleaning.

Modern machines are also equipped with advanced control systems to indicate and automatically adjust the weight of the pieces and to regulate the operating speed, and with a dusting device by which a suitable amount of flour is dusted on to the conveyor belt to prevent the newly cut dough pieces from sticking to the web.

6.3   DIVIDING PRINCIPLE

The weight of a dough piece of a given size is a function of dough density, which in turn is influenced by such factors as time, temperature, yeast activity, etc [2]. It is necessary to control all the process conditions to ensure a uniform density (homogeneous structure). So long as the dough density is kept constant, the weight as well as the volume of the dough pieces will be uniform. Fig. (6.2) is a schematic diagram of the volumetric dividing principle.

The dividing process consists of suction of the fresh dough, volumetric scaling, and discharging. The sucking action of the fresh dough always takes place when a scaled dough piece is being discharged.

Fig. (6.2a) shows the division box (6) with the die (7) going up (returning) to its top position, as the dough is drawn from the hopper (4) to the underlying dough chamber (9). The knife (2) and the ram (1) begin to move leftward horizontally. As the knife (2) moves further, it cuts off a portion of dough at the bottom of the hopper (4) and blocks the passage between the hopper and the dough chamber (Fig. 6.2b). The ram (1) then pushes the several dough pieces into the division box (6) (pockets), forcing the die (7) out. In some machines, the die and the discharge block are controlled by the same mechanism. When the preset volume is reached (this is, the limit of the preset stroke of the die) the die (7) stops.

Next, the filled division box (6) takes the charge of dough downwards (Fig. 6.2c). By means of shearing, the charge in the pockets of the division box (6) is cut off from the dough mass of the dough chamber as the pockets move past the shear edge of the front plate, while the ram (1) is moved to the right, and a vacuum is formed in the emptied space of the chamber (9). When the knife (2) moves to the right, the passage is opened to allow the dough to be drawn into the chamber (9) by vacuum suction and gravity. That is why the chamber is referred to as suction chamber.

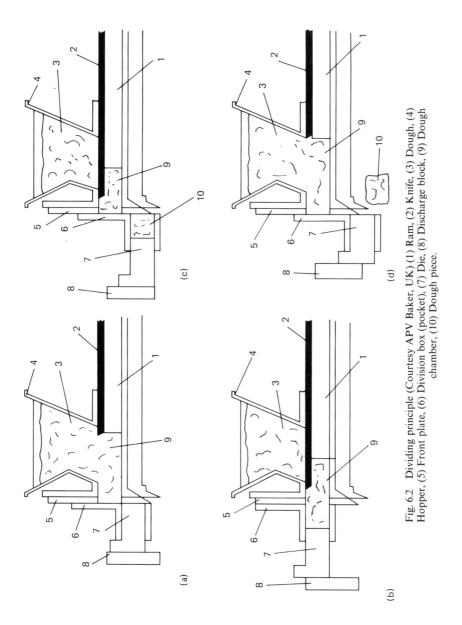

(a)

(b)

(c)

(d)

Fig. 6.2   Dividing principle (Courtesy APV Baker, UK) (1) Ram, (2) Knife, (3) Dough, (4) Hopper, (5) Front plate, (6) Division box (pocket), (7) Die, (8) Discharge block, (9) Dough chamber, (10) Dough piece.

When the division box (6) arrives at its lowest position (Fig. 6.2d) the discharge block (8) pushes the die forward so that the dough piece is discharged onto the primary conveyor which leads to a rounder. As the newly cut bread doughs used are very soft and sticky, it may sometimes not be possible to separate them from the edges of the division box and the die completely. To prevent any residual material, a plate is placed in front of the chamber against the division box to scrape against the die and box surfaces as they are returning to their upper position. By then, more dough has been sucked down from the hopper and into the suction chamber for the next cycle of division.

6.4 TRANSMISSION

From the variable speed rotor (1) (Fig. 6.3) through pulleys (2) and (3), chain drives (4) and (9), rotation is transmitted to shaft II by which the primary (receiving) conveyor (28) is made to deliver the divided dough pieces to the rounder. The gear (26), by means of idle gear wheel (25), actuates the gear (6) and the crankshaft III, on which three cranks (5), (7) and (24) are mounted.

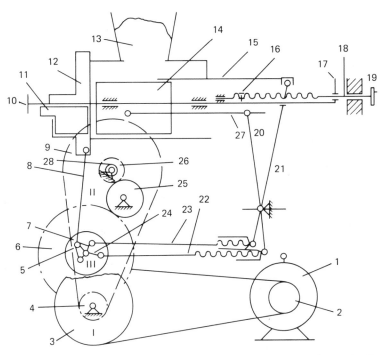

Fig. 6.3 Transmission system of a typical divider (Courtesy of Xin Xiang City Food Machinery Plant, China) (1) Motor, (2), (3) Pulleys, (4) Chain gear, (5) Crank, (6) Gear, (7) Crank, (8) Connection rod, (9) Chain gear, (10) Discharge block, (11) Die, (12) Division box, (13) Hopper, (14) Ram, (15) Knife, (16) Scaling indicator, (17) Discharge control block, (18) Screw lever, (19) Handwheel, (20), (21) Rocking levers, (22) (23) Connecting rods, (24) Crank, (25), (26), Gears, (27) Connecting rod, (28) Primary conveyor. (I), (II), (III) Shafts.

    The chain gear (9) is linked to shaft II by a pin which ensures that the machine stops
as soon as an overload happens.
    From crank (24), through connection rod (22), rocking lever (20), and connecting
rod (27), the ram (14) is driven reciprocally in a horizontal plane. This action allows
the dough to move downward into the chamber and pushes it further into the division
box. The amount of dough allowed into the chamber is dependent upon the travel
length of the ram (14) which can be adjusted by means of handwheel (19) which
regulates the length of connecting rod (22).
    From crank (7), through connecting rod (23) and rocking lever (21) the knife (15)
reciprocates horizontally to open and shut the passage and cut the dough at the
hopper outlet. The travel distance of the knife is controlled by the length of the
connecting rod (23). Simultaneously, the rocking lever (21) causes the discharge
control block (17) and the discharge block (10) to reciprocate horizontally and
intermittently so that the die pushes the divided dough piece out of the division box
(pocket). The travelling distance of the die (11) can be adjusted by turning handwheel
(19), via screw (18) to change the position of the discharge control block (17).
    From crank (5) and connecting rod (8) the division box (12) is made to reciprocate
vertically to allow the dough into or out of the pocket during the dividing operation.
    The transmission route of the divider is shown in Scheme 6.1.

$$\text{Motor 1} \rightarrow \frac{\text{Pulley 2}}{\text{Pulley 3}} \rightarrow \text{Shaft I} \rightarrow \frac{\text{Chain gear 4}}{\text{Chain gear 9}} \rightarrow \text{Shaft II}$$

$$\text{Shaft II} \rightarrow \begin{cases} \text{Primary conveyor 28} \\[2mm] \dfrac{\text{Gear wheel 26}}{\text{Gear wheel 25}} \rightarrow \dfrac{\text{Gear wheel 25}}{\text{Gear wheel 6}} \rightarrow \text{Shaft III} \end{cases}$$

$$\text{Shaft III} \rightarrow \begin{cases} \text{Crank 24} \rightarrow \text{Connecting rod 22} \rightarrow \text{Rocking lever 20} \rightarrow \text{Connecting rod 27} \\[2mm] \text{Crank 7} \rightarrow \text{Connecting rod 23} \rightarrow \text{Rocking lever 21} \rightarrow \begin{cases} \text{Knife 15 (reciprocating horizontally)} \\[2mm] \text{Discharge block 17, 10} \end{cases} \\[2mm] \text{Crank 5} \rightarrow \text{Connecting rod 8} \rightarrow \text{Division box 12 (reciprocating vertically)} \end{cases}$$

Connecting rod 27 → Ram 14 (reciprocating horizontally)

Discharge block 17, 10 → Die 11 (reciprocating horizontally)

Scheme 6.1.   Transmission scheme for a divider.

6.5   OPERATION AND MAINTENANCE

6.5.1   Lubrication

Before each start the lubrication systems should be checked. The gear drives, bearings, and other moving parts which are not in contact with the dough are lubricated with the usual petroleum oil and greases; drives at the lower part of the machine can be greased, while other parts in contact with the dough must be lubricated with a technical white grade mineral oil with a viscosity of 85 to 120 at 212°F and a flash point 662°F. To restrict the oil content to under 0.15%, the correct setting is that at which the least amount of oil is used to secure the proper function of the divider [1].

6.5.2   Weight control

If dough pieces of unequal size are produced, the pin at the rear of the ram should be adjusted gradually from right to left by trial and error until the correct uniform size is achieved.

Be careful to avoid moving the pin to the extreme right end, where it would cause a sharp increase in loading to shorten the machine life by excessive suction and compression of the dough.

If the pieces are uniform, but with a large error, this can be regulated by a turn of the handwheel (19) (Fig. 6.3) to adjust the limit position of the die (11) via the screw lever (18) which enables one to obtain the correct volume in the division box (pocket). The rotation (clockwise and anticlockwise) of the scaling indicator (16) shows the increase and decrease of the dough piece weight.

The proper clearances between the knife and the ram as well as between the division box and the front plate are very important for the correct operation of the machine. Insufficient clearance will cause severe friction and overheating which will have a bad effect on the dough and extra demands on the motor. In contrast, excessive clearance will lead to dough leakage. Therefore the maintenance instructions must be strictly followed.

6.5.3   Cleaning

The machine should be cleaned whenever the machine is shut down for more than an hour, to avoid caking of the retained dough. Fortunately, the modern divider is equipped with a hinged hopper which allows easy access to the dividing assembly. The ram, knife, and dies are easily removed for cleaning. Attention should be paid to the identifying marks on the dies, for they are usually not interchangeable. Only a hardwood or plastic scraper should be used to remove the excess dough from the parts, which should then be washed with water and soda or a detergent. Lastly, the parts should be rinsed, dried, and covered with a thin film of divider oil.

7

Rounders

7.1 INTRODUCTION

The dough pieces transmitted from a divider are irregularly-shaped, with a sticky fresh-cut skin and uneven structure in which the gluten network is disoriented. The function of the rounder is to ball up the dough pieces as well as to give them a thin, smooth, and dense skin with a reoriented gluten structure, which is necessary for the later bread-making process.

The moulders used today are classified into three types: umbrella, bowl, and drum type, which all work on much the same principle. They differ in the shape of their rotatable body, which has the appearance of an umbrella, a bowl (cone), or a drum respectively. The irregular dough pieces move up or down on the revolving body surface, guided by a stationary spiral track, and are well-rounded with a relatively dried exterior and the desired structure. Fig. (7.1) illustrates modern rounders.

7.2 CONSTRUCTION

The rounders will be discussed by taking the umbrella-type rounder as an example (Fig. 7.1a and Fig. 7.2).

The rounder essentially consists of a rotary working surface, a stationary spiral track (also named dough race, or dough trough) wrapped round the conical surface of the body, a fan or flour-dusting device, a discharge cone, a chute, a frame, and a power and transmission system.

To support the working surface and to stabilize the motion, the stand and frame of the machine are usually made of cast iron. The motor is mounted at the bottom.

7.2.1 Working surface

Above the transmission case is a rotatable conical working surface resembling an umbrella, which is supported and actuated by a vertical shaft driven by a worm gear (Fig. 7.2).

The outside surface of the hard-wearing cast iron cones is differently treated by different manufacturers. Some are smooth, some are grooved, and some are corrugated vertically or horizontally in various sizes and designs to increase the friction

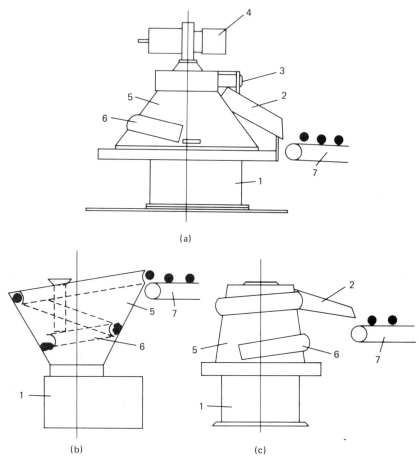

Fig. 7.1   Three types of rounders (Courtesy of APV Baker, UK) (a) Umbrella, (b) Bowl, (c) Drum. (1) Stand, (2) Chute, (3) Discharge cone, (4) Fan, (5) Rotatable working surface, (6) Dough trough, (7) Conveyor to intermediate prover.

between the dough piece and the revolving surface. Some are waxed or coated with Teflon to reduce sticking.

7.2.2   Dough trough

The stationary dough trough, which is also named 'race' or 'track' or 'channel', is supported against the revolving cone surface, spirals upward from the bottom outer edge to the central apex of the cone.

A series of standard troughs is available for a wide range of bread sizes, for example, the APV Baker series for bread weights of 50, 100, 200, 250, 300, 400, 620, 800, 1600 g. The capacity of the machine is dependent upon the size of the dough pieces being produced. The revolving speed of a working surface is usually constant, and is suitable for a series of standard dough troughs. Table (7.1) shows specifications of some modern rounders.

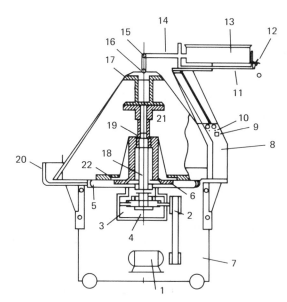

Fig. 7.2   Construction and transmission of an umbrella type rounder (Courtesy of Xin Xiang City Food Machinery Plant, China) (1) Motor, (2) Pulley and belt, (3) Worm gear, (4) Worm gearbox, (5) Spindle support, (6) Bearing seat, (7) Frame, (8) Support for dusting device, (9) Adjusting screw, (10) Screw, (11) Control plate, (12) Butterfly bolt, (13) Dusting box, (14) Shaft, (15) Pulling rod, (16) Cap, (17) Revolving surface, (18) Spindle, (19) Connecting plate, (20) Baffle, (21) Flange connection, (22) Bearing.

7.2.3   Fan and dusting devices

Some machines are equipped with a dusting device (Figure 7.2) to supply a controlled amount of dusting flour to prevent the dough from sticking either on the revolving conical surface or on the spiral dough track, since the surface of a newly divided dough piece is always ready to adhere to a metal surface. The dusting flour also improves the

Table 7.1   Specifications of typical rounders†

No.	1	2	3	4	5
Capacity (pieces/h)	max. 9000	max. 4800	4800	7200	6000
Weight scale (g/piece)	400, 620, 800, 1600 (finished bread)		30, 230 (dough)	30, 65 (dough)	400, 500 (dough)
Motor	Main 5.5 kW Fan 0.75 kW	Main 3.0 kW Fan 0.75 kW	Main 0.75 kW Dusting 75 W	Main 1.5 kW Dusting 75 W	
Dimension (mm)	Max. 1510 × 1510 × 1703		920 × 920 × 1310	1400 × 1400 × 1800	
Net weight			350 kg	1200 kg	

†No. 1 and 2 by courtesy of APV Baker, UK.
No. 3, 4, and 5 by courtesy of Tokyo Maruichi Shoji Co., Ltd, Japan

ability of the forces applied to the dough to work more efficiently. Some machines, however, have a fan mounted at the apex of the cone (Fig. 7.1a) or a forced air unit. Air blows down on to the trough, and a skin forms on the dough, which can prevent sticking and is helpful for rounding. These blown air systems save the cost of dusting flour. For all normal doughs, about 20 m^3/min (700 ft^3/min) of air is needed, while 42 m^3/min (1500 ft^3/min) is required for sticky doughs (APV Baker).

The discharge cone (Fig. 7.1a) is fitted at the top part of the machine near the outlet of the dough trough to steady the rounded pieces for discharge to a conveyor leading to the first prover.

7.3 ROUNDING PRINCIPLE

Fig. (7.3) illustrates the rounding principle of a typical umbrella type rounder. The dough piece transmitted from a divider arrives at the entrance of the rounder surface. Friction immediately engages it with the revolving conical surface and carries it around and upward into the spiral trough where compound forces act on it. The forces are friction between the dough piece and the trough, friction between the dough piece and the cone surface, centrifugal force, and gravity. The complexity of the compound force is due to the shapes of the revolving surface and of the trough and to

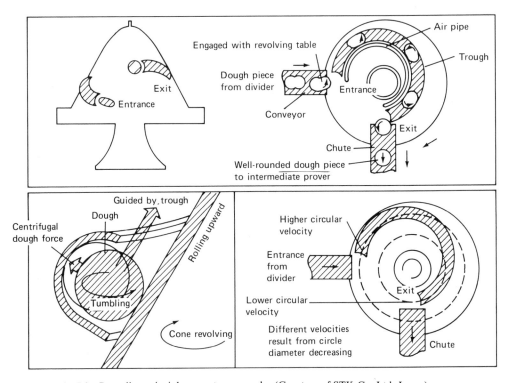

Fig. 7.3   Rounding principle, cone type rounder (Courtesy of STK Co. Ltd, Japan).

their relative motion. The irregularly shaped dough piece lies between the internal curved surface of the stationary spiral trough and the curved external surface of the rotating cone. The dough piece moves upward and also moves, three-dimensionally, around its own centre. There is also a slight sliding action. One may say that the dough piece is tumbled and rolled from start to finish of the rounding action.

The finished dough piece is substantially spherical, with a smooth continuous skin and a dense uniform texture. The gluten network has also been reoriented. Steadied by the discharge cone, the rounded dough pieces fall onto the belt leading to the intermediate prover.

As shown in Fig. 7.3, the dough piece enters the rounder at the largest diameter of the cone, and its initial movement is therefore faster than at the outlet of the trough (the top of the 'umbrella', with a much smaller diameter). The distance between two pieces therefore becomes smaller and smaller from the bottom to the top of the revolving cone. Uneven feeding or any abnormal conditions would lead to 'doubles'. A device to shunt aside oversize dough pieces (doubles) is mounted at the outlet of the chute, and the expelled dough is re-used.

As contrasted to the umbrella type rounder, the working surface of a bowl type rounder is an inverted cone. Inside the cone is a stationary spiral dough trough from the lower to the upper part of the bowl (Fig. 7.1b). A special tube-like hopper is vertically supported inside the bowl. The divided dough pieces fed from the hopper intake drop to the lower part of the bowl and are immediately engaged into the spiral trough by the revolving bowl surface. Under a similar compound force action, as in the 'umbrella' rounder, the dough pieces travel upward as they are tumbled and rolled in the spiral trough and are rounded into dense balls until they are discharged onto a belt leading to the intermediate prover. Since the dough enters the rounder at the smaller diameter of the bowl, its initial movement is slower than at any other part of the bowl. This means that the distance between two neighbouring dough pieces becomes longer and longer during their progress to the top of the bowl. Therefore no 'doubles' will form. However, this advantage does not make the bowl type more popular than the umbrella type rounder. The present writer finds that the umbrella is the most popular one among the three types of rounder because of its convenience and the better rounding quality achieved, which is similar to that of the hand-made product.

The drum type rounder (Fig. 7.1c) has an almost cylindrical working surface. The dough pieces therefore travel at a more uniform rate in this type of machine than in the other two varieties. Furthermore, the smaller slope of the drum type rounder makes it possible to save floor space. Because of these advantages, the drum type rounder should be the most popular, but this is far from being the case. In the bread plants of countries which the writer has visited, the umbrella type rounder is the most widely used.

7.4 TRANSMISSION

The transmission system of the rounders is simple compared with those of other food machinery. As shown in Scheme 7.1, the power is transmitted from a motor (1) to a worm gear reducer through a belt and pulley gear (2). Actuated by the worm wheel, the vertical shaft (18) revolves so that the conical working surface (17) revolves along with it.

In some rounders, the dusting device is driven by a separate electric motor mounted at the top of the machine. Fig. 7.2 shows an eccentric hole on the top cap (16) of the revolving cone. Between the hole and the pulling rod (15) is a spherical connection. By means of the pulling rod (15), the shaft (14) of the flour dusting box (13) swings radially to distribute the flour evenly onto the spiral trough to prevent any sticking. When the machine stops, the butterfly bolt (12) should be loosened and the flour outlet of the box should be shut by the control plate (11). The transmission route of the rounder can be expressed as follows:

Motor 1 → belt and pully 2 → worm gear 3 → spindle 18

Spindle 18 → { Conical surface 17 (revolving) → dough pieces being rounded

Conical surface 17 → hole 16 → pulling rod 15 → shaft 14
(swing: dusting flour)

Scheme 7.1.   Transmission scheme for a rounder.

7.5   OPERATION

7.5.1   Dough piece shape control
To prevent the formation of 'pills' or small pieces of dough which are pinched off between the trough and the revolving conical surface during operation, the gap must be correctly controlled. For a better rounding effect, the edge of the stationary trough should be as close as possible to the rotating cone surface, but without creating friction. However, there is always a tendency for a film of dough to build up as the rounder operates. The distance of the trough edge from the rotating surface of the cone is therefore gradually reduced as the operation continues, so that a periodic readjustment is needed.

7.5.2   Dusting flour control
Too much dusting flour will lead to cores in the finished bread, while too little dusting will lead to sticking and to failing of the operation. To adjust the amount of dusting flour, the control plate (11) can be used to regulate the opening of the sifter at the bottom of the flour box (Fig. 7.2).

7.5.3   Cleaning
It seems inevitable that some bits of dough will adhere on either the surface of the cone or the inside of the dough trough. This leads to deterioration in both the rounding process and the quality of finished loaves. Therefore it is necessary to carefully clean

all the parts in contact with the dough. A hardwood or plastic scraper, which will not score the surfaces, is used.

The rate at which the divided dough pieces are fed should be controlled by adjustment of the divider speed, since the speed of the rounder may be constant. Too fast a feed will cause 'doubles', while too low a speed will reduce the production rate.

Part IV

Biscuit-making machines

8

Sheeting machines

From the mechanical point of view, sheeters, gauge rolls, and laminators are all devices for reducing the dough thickness by compressing and gauging the dough mass into a sheet by means of chain rolls. They are used in the biscuit production lines, and they vary so much that we can provide only a general instruction to them for the buyers, users, and students.

8.1 SHEETING PROCESS AND SOME PARAMETERS

8.1.1 Ideal sheeting process

During sheeting, the degree of plastic deformation of the dough depends on many factors. For the convenience of analysis, it would be better to begin the study with the ideal sheeting process, which generally satisfies the following conditions.

(a) Equal diameter and speed of each pair of rolls, driven at the same speed.
(b) Uniform motion of the dough sheet, as in a web.
(c) Absence of forces on the dough sheet, apart from roll pressure and gravity.
(d) Uniform and continuous mechanical properties.
(e) Constant volume of the dough sheet.

It is obvious that there are differences between the ideal and the actual sheeting process. But these can be corrected and perfected by means of practical experiments from which some functional results can be achieved.

8.1.2 Contact angle

In the sheeting process, the deformation takes place as the dough comes into contact with the rolls. Fig. 8.1 shows the deformation area where D is the sheeting roll diameter, $D = 2R$; h_0 is the thickness of the sheet before sheeting; h_1 is the sheet thickness after sheeting; Δh is the absolute reduction of the sheet thickness; L_0 is a part length before sheeting; L_1 is the length of the part after sheeting; B_0 is the width of a part before sheeting; B_1 is the width of the part after sheeting; ΔB is the absolute width increment of the part, $\Delta B = B_1 - B_0$; α is contact angle; and e is the gap between two rolls, theoretically, $e = h_1$.

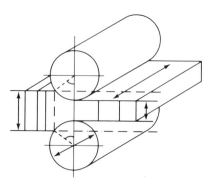

Fig. 8.1   Sheet deformation area.

The contact angle, α, is the centre angle of the roll formed by the deformation area to the roll centre. From Fig. 8.2, it can be expressed as

$$\cos \alpha = \frac{R - \frac{1}{2}\Delta h}{R} \quad \text{or} \quad \frac{\Delta h}{2R} = 1 - \cos \alpha$$

$$\frac{\Delta h}{2R} = 2 \sin^2 \frac{\alpha}{2} \qquad \sin \frac{\alpha}{2} = \frac{1}{2}\left(\frac{\Delta h}{R}\right)^{1/2} \tag{8.1}$$

For a sheeter, as its roll radius is known, Δh can be directly measured, and the contact angle during sheeting can be calculated from equation (8.1).

8.1.3   Initial condition and sheeting roll diameter

In the contact area, A is a point on the dough that is to be sheeted (Fig. 8.3), just before it is introduced into the gap between the pair of rolls. P_n is the normal pressure exerted by the roll on point A, P_f is the friction force imparted by the roll to point A, ϕ is the friction angle, and α is the angle of introduction.

P_n and P_f can be resolved into the horizontal force P_{nx}, P_{fx} and the vertical force P_{ny}, P_{fy}. From this, the correct initial condition is that the effect of the friction force should be larger than the effect of the normal pressure on the dough in a horizontal direction. That is,

$$P_{fx} > P_{nx}. \tag{8.2}$$

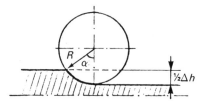

Fig. 8.2   Contact angle.

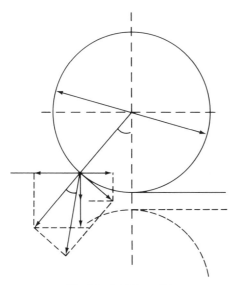

Fig. 8.3   Initial condition.

From equation (8.2),

$$P_n \tan \phi \cos \alpha > P_n \sin \alpha \qquad (8.3)$$

From equation (8.3),

$$\tan \phi > \tan \alpha \qquad (8.4)$$

or

$$f > \tan \alpha \qquad (8.5)$$

where f is the friction factor,

$$\frac{P_f}{P_n} = f.$$

Equation (8.4) is the initial condition for correct sheeting operation. Only under this condition can the dough be smoothly and continuously introduced into the gap and be sheeted by the rolls.

From equation (8.4), the diameter of the sheeting rolls can be derived as the basis of sheeting roll design. From Fig. 8.2,

$$\tan \alpha = \left\{ \frac{(D/2)^2 - [(D - \Delta h)/2]^2}{(D - \Delta h)/2} \right\}^{1/2} = \frac{(D^2 - (D - \Delta h)^2)^{1/2}}{D - \Delta h} \qquad (8.6)$$

Substituting equation (8.6) for $\tan \alpha$ in equation (3.4),

$$\tan \alpha > \frac{(D^2 - (D - \Delta h)^2)^{1/2}}{D - \Delta h} \qquad (8.7)$$

From Equation (8.7), the functional relation between the sheeting roll diameter, the absolute thickness reduction, and the friction angle can be derived as:

$$D > \frac{2\,\Delta h}{\sin^2 \phi} \qquad (8.8)$$

Equation (8.8) shows that the diameter of the sheeting roll and the absolute reduction of the sheet thickness should be properly selected. The larger that D is, or the smaller that Δh is, the easier the sheeting operation will be. Reversely, too large a Δh will result in too great a roll diameter. Theoretically, a single pair of rolls should not reduce the sheet thickness by more than 50% for a reasonable roll size and machine dimensions. To minimize the stress introduced by the gauge rolls, it is common to make the thickness reduction more at the pre-sheeting stage, but less at the gauging stage. That is why the diameter of the pre-sheeter rolls is larger than that of the gauging rolls. Equation (8.8) also explains why the sheeting operation cannot be carried out in one or two stages. Food technology textbooks give technological reasons for this. Equation (8.8) explains the mechanics of the matter.

8.2 SHEETERS

8.2.1 Two-roll pre-sheeters

Machines in this group are used for pre-sheeting the dough into a rough thick sheet which is transferred by a conveyor to feed a laminator or gauge rolls at the beginning of a baking line.

The machine is generally composed of a portable hopper with open base and intermediate chute, and two rotary grooved rolls. The machine frame is commonly a substantial fabricated one connected by stretchers and covered by end plates to form a rigid box section structure. Some machines are equipped with a stirrer in the dough hopper to prevent 'bridging' over the rolls, which often happens with soft doughs. Fig. 8.4 shows the position of the sheeters and the forming machines such as the rotary cutter in a biscuit production line. Fig. 8.5 shows the sheeting process carried out by a two-roll pre-sheeter.

The diameter of each roll is about 400 mm, with the width ranging from 500 to 1500 mm to match the width for the next stage equipment. Each roll is fitted with a scraper to clean the roll surface.

As the dough is discharged into the hopper, it is dragged by the contra-rotating rolls into the 'nip' between them, where the dough is roughly metered and compressed out as a thick dough sheet which then drops onto an underlying conveyor, commonly made of flat woven cotton, by which the thick dough sheet is fed to the next equipment hopper at a rate set by the baking line, most often depending upon the speed of the oven band. Therefore the machine is always equipped with either a variable speed motor or other variable speed unit.

The output of the machine is dependent upon the roll speed and the gap between the rolls. The gap, commonly ranging from 15 to 45 mm, can be adjusted by operating a handwheel or pushing a button servo system connected to the facia panel.

Table 8.1 shows specifications of this group of sheeters.

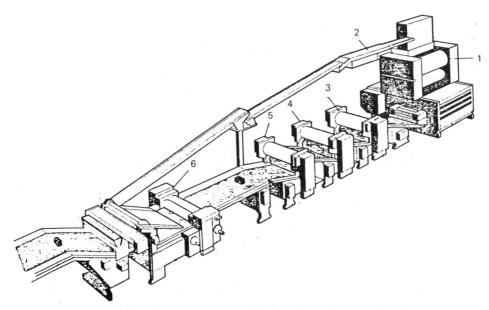

Fig. 8.4   Typical biscuit production line (Courtesy of APV Baker, UK) (1) Vertical laminator (or pre-sheeter), (2) Scrap return, (3), (4), (5) Gauge rolls, (6) Biscuit-cutting machine

8.2.2   Three-roll sheeters

Sheeters of this type have a dough hopper with an adjustable division plate so that the returned scrap dough is at the bottom of the later sheet, which reduces adherence to the woven cotton web of the dough sheet conveyor; the three rotating rolls are arranged as shown in Fig. 8.6.

The rolls are made of heavy duty cast iron. The forcing roll (1) and top roll (2) are either both heavily grooved across their length, or only roll (1) is grooved. They form the first pair of rolls, the pre-sheeter, while the bottom roll (3) is smooth and flanged to

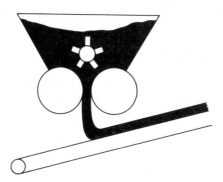

Fig. 8.5   Two-roll pre-sheeter.

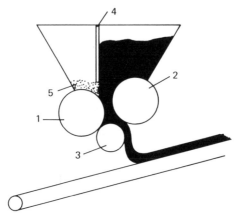

Fig. 8.6   Three-roll sheeter (front discharge type) (1) Forcing roll, (2) Top roll, (3) Bottom roll, (4) Division plate, (5) Return scrap.

limit the increment in the width direction of the sheet during sheeting. Rolls (2) and (3) form the second pair, the gauge rolls. The three rolls are commonly mounted between the machine sideframes and are supported by either double row spherical roller or deep groove ball bearings. Each roll is equipped with a scraper made of spring steel. The gap between rolls (2) and (3) is adjusted by altering the position of roll (2) rather than roll (1), commonly via eccentric bearing housing and screwjacks or wormwheels.

The lengths of the roll are designed to suit the width of the next equipment, ranging from 560 mm to 1500 mm, while the diameter of rolls (1) and (2) is about 400 mm, and that of roll (3) about 300 mm. The speed of rolls (1) and (2) is designed to give the fresh and the returned scrap dough a suitable compaction.

There are two types of dough sheet discharge, front and back. Fig. 8.6 shows the front type which is suitable for weak and short dough where support is needed as the sheet leaves the sheeter [11]. The back type is preferred for all extensible doughs, since their gluten networks allow them to withstand the tension (Fig. 8.7).

Below the sheeting rolls is a woven cotton or food quality synthetic fabric web which is generally supported by a stainless steel table. To ensure proper feeding for the next stage equipment, some installations are equipped with web tracking and web

Table 8.1   The specifications of two-roll presheeters (APV Baker, UK)

Model	1200 mm width unit	1000 mm width unit
Output (at gap 25 mm)	6000 kg/h	5000 kg/h
Roll periphery speed	3 m/min	3 m/min
Gap range (mm)	15 to 45	15 to 45
Electric motor (kW)	3	3
Net weight (kg)	3000	2500

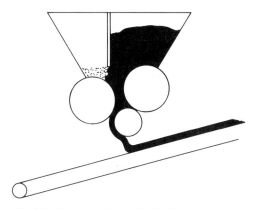

Fig. 8.7   Three-roll sheeter (back discharge type).

tension. For modern machines, the web speed is regulated in relation to the roll speed by a potentiometer mounted on the machine facia with access via selector switch.

As well as the variable speed unit, modern machines are provided with facilities to monitor the hopper level and the chamber pressure, since they are factors which influence the sheet flow rate.

Table 8.2 shows specifications of some modern three-roll sheeters.

8.2.3   Four-roll sheeter

Sheeters of this type are also used as a pre-sheeter and feeder to the gauge rolls or laminator at the head of a biscuit production line. But they produce smoother, more even, and more precisely metered dough sheet than the previously mentioned sheeters,

Table 8.2   Specifications of some modern three-roll sheeters (APV Baker, UK)

Model	1	2
Roll drive motor (kW)	7.5 max.	7.5 max.
Web drive motor (kW)	1.1 max.	1.1 max.
Roll adjustment (mm)	3 to 16	7 to 20
Forcing roll diameter (mm)	420	420
Top roll diameter (mm)	420	420
Bottom roll diameter (mm)	300 flanged	230 flanged
Sheet width (mm)	812, 1016, 1220, 1270	800
Dimensions (mm)	1480 × 1532 × 1650 1480 × 1736 × 1650 1480 × 1940 × 1650 1480 × 1990 × 1650	2185 × 1852 × 972
Net weight (kg)	4253, 4877, 5384, 5420	4000

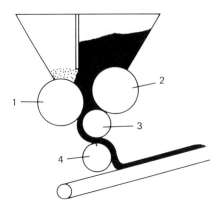

Fig. 8.8   Four-roll sheeter. (1), (2) Sheeter rolls, (3), (4) Gauge rolls.

since it comprises four rotating rolls arranged as shown in Fig. 8.8. Table 8.3 shows specifications of some machines of this type.

Rolls (1), (2), and (3) form a three-roll sheeter which discharges the dough sheet in the back type to the gauge roll unit. Rolls (3) and (4) gauge the sheet into preset thickness. Usually, the forcing roll (1) is grooved, the top roll (2) fluted, and rolls (3) and (4) are matt finished.

The sheeter roll gap between rolls (2) and (3), ranging from 5 to 20 mm, can be regulated by changing the position of roll (2) via an eccentric bearing housing and screwjacks or by other mechanisms. The gauge roll gap between rolls (3) and (4) is commonly adjusted by changing the position of roll (4) by means of eccentric bearing housing and screwjacks.

Table 8.3   Specifications of some four-roll combination sheeters
(APV Baker)

	Sheeter roll		Gauge roll	
Roll diameter (mm)				
	420	420	230 flanged	230
Roll adjustment	5–15 mm		3–23 mm	
Roll drive motor		7.5 kW max.		
Web drive motor		1.1 kW max.		
Dimensions (mm)	1560 × 2101 × 2005 for 812 mm plant 1560 × 2305 × 2005 for 1016 mm plant 1560 × 2509 × 2005 for 1220 mm plant 1560 × 2559 × 2005 for 1270 mm plant			
Net weight (kg)	5470	5977	6484	6563

To synchronize with the speed of the baking line, the rolls and the double drums of the web are driven by chains and gearboxes coupled to the corresponding variable speed units.

The hopper resembles that of the three-roll sheeter, which consists of a scraper return compartment with an adjustable division plate. All components within and adjacent to the product area are either stainless steel or are nickel-plated.

Machines of this group are also provided with facilities for web tracking and web tension, as well as an operator facia control to adjust the speed and web and to regulate the gap between rolls (2) and (3), (3) and (4). The option ratio of the forcing roll gap between rolls (1) and (2) is maintained.

8.3 GAUGE ROLLS

The gauge rolls compress the thick primary dough sheet, fed from the previously mentioned sheeters or laminators, to the thickness required by the next stage machinery such as a docker or rotary cutter.

A gauge roll unit comprises two large smooth rolls made of heavy duty cast iron covered with high chromium stainless steel and mounted one above the other between thick plate sideframes via double row spherical roller bearings. The roll diameter ranges from 230 mm to 400 mm, and the length matches the width of the oven band, from 560 mm to 1500 mm.

The gap between the two rolls is theoretically the thickness of the cut piece products that are to be formed by a docker or a rotary cutter. It has been commonly adjusted by changing the top roll position, but modern machines are regulated by raising or lowering the bottom roll, in preference to adjustment of the flanged top roll, via a handwheel with integral indicator. In this way, the working clearances are taken up by gravity in both the static and running conditions and ensure that the adjustable roll does not rise and fall as the hardness of the dough sheet varies. This is an APV Baker patented feature for maintaining constant dough sheet thickness.

The rolls are generally provided with aluminium bronze scrapers. The one for the top roll is commonly fixed; that for the bottom roll is adjustable to serve as the nosepiece of the delivery table which supports the dough sheet web.

Like other sheeters, the speed of the gauge rolls and the delivery web are adjustable to synchronize with the succeeding equipment. For some modern machines, the drive between the top and bottom rolls is designed to allow the relative surface speeds to be readily altered. The tendency of sticky dough to ride up with the top roll can be overcome by an arrangement to alter the roll differential (APV Baker, UK).

The number of gauge rolls is dependent upon the properties of the dough, the thickness of the later products, and the type of the pre-sheeter used. Too sharp a reduction on the dough sheet thickness would introduce stress in the dough which would deform the products, while too slow a thickness reduction would cost more for equipment and energy consumption. For most products, two or three sets of gauge rolls are employed before a docker or rotary cutter, while four sets of gauge rolls are needed for very thin products of very elastic doughs. Fig. 8.9 illustrates a gauging process performed by two sets of gauging rolls.

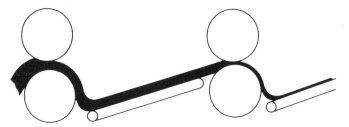

Fig. 8.9   Sheeting process with two sets of gauge rolls.

The gap between the first pair of rolls generally ranges from 0 to 15 mm, and the final gap from 0 to 5 mm. As the gap becomes smaller, the sheet becomes thinner, and the speed of the conveyor web should be increased after each gauge roll unit. Otherwise, the sheet will pile up. Theoretically, the speed ratio of the feed web to the take-away web is the inverse thickness ratio of the feed sheet to the gauged sheet.

Table 8.4 shows specifications of some modern gauge rolls.

To assist separation of the dough sheet and the top roll, a flour spreader or an air blast is employed in a biscuit production line. There is a tendency for the air blast to supersede the spreader. In the flour spreader (sprinkler) system, flour is used as a release agent, while the blast is used to blow air onto the sheet to form a dried 'skin' to prevent sticking.

The air blast forces air from a row of holes located along a tube positioned across the dough sheet. This unit is equipped with a motorized fan connected to the blowing tube which can be rotated to alter the angle of incidence of air on the dough sheet, as a heater is incorporated between the fan and the tube.

Table 8.4   Specifications of some modern gauge roll units (APV Baker)

Model	1	2
Roll diameter	230 mm	300 mm
Roll gap	0 to 3 mm on final stage 0 to 11 mm on other stages	0 to 3 mm on final stage 0 to 15 mm on other stages
Roll drive	3.0 kW (standard machine) max. 4.0 kW (high speed) max.	11 kW max.
Web drive	0.55 kW (standard machine) 0.75 kW (high speed)	1.5 kW max.
Dimensions (mm)	1500 × 1706 × 1350 (812 machine) 1500 × 1910 × 1350 (1060 machine)	1286 × (plant width 825) × 1470
Net weight (kg)	1750 (800 mm wide), 2030 (1000 mm wide), 2600 (812 mm wide) 2987 (1016 mm wide), 3340 (1220 mm wide), 3440 (1270 mm wide)	

8.4 LAMINATORS

For puff biscuits, the dough sheet should be made up from many extremely thin sheets
sandwiched with fat. This kind of sheet used to be hand-made by highly skilled people
before the invention of modern laminators, which accept the bulk dough or a rough
thick dough sheet from a two-roll presheeter and make it into the required overall
thickness.

In terms of sheet flow direction, modern laminators are of two types, vertical and
horizontal. In terms of the laminating process, they are continuous layering laminators or cut-sheet layering laminators.

In terms of the fat filling method, they are of dusting (sprinkling) and extruding
type. The dusting type is the more widely used.

8.4.1 Laminators with a filling duster

Most of these laminators are composed of a three-roll sheeter, one or two or three sets
of gauging rolls, a filling duster, a release agent (flour) duster, and a continuous
layering section or a cut sheet layering section. If the gauging rolls are arranged
vertically one above the other, the dough sheet moves horizontally as it passes
through the gauging rolls, so that the machine is called a horizontal laminator. If the
gauging rolls are arranged in horizontal parallel, the dough sheet moves vertically as it
passes through the gauge rolls, so that it is called a vertical laminator.

The operating principles of the two groups of laminator are similar, but the vertical
machines are more popular since they require less floor space.

To save floor space, some horizontal laminators are designed with two stacked
sections one above the other instead of in succession at one level. They are called
stacked horizontal laminators.

Because of their similarities, only the vertical laminators will be discussed in this
section.

8.4.1.1 *Vertical laminators with a continuous layering unit*

As previously mentioned, there are two types of layering unit for laminators. Fig. 8.10
shows a typical vertical laminator with a continuous layering unit. Fig. 8.11 shows a
laminator with a cut sheet layering unit.

As shown in Fig. 8.10, the upper part of the machine comprises a three-roll sheeter
under a dough hopper, by which the dough is formed into a thicker sheeter. To
minimize the tension stress in the dough, an intermediate horizontal web conveyor is
placed under the three-roll sheeter to allow the sheet to rest for a while before the two
pairs of gauge rolls reduce it to the required thickness. After gauging, another
horizontal but variable speed web conveyor is provided to relieve some of the tension
in the sheet. As the right end of this conveyor, the sheet is dusted with flour by a duster
(sprinkler) to prevent the sheet sticking to the conveyor of the continuous layering
section, since the sheet is upside down as it comes back under itself in the next stage.

The continuous layering process is performed by a carriage which reciprocates
across the discharge web, running at right angles to the dough sheet as it feeds the fat/
flour mixture onto the sheet surface. The layering stroke length matches the width of
the next stage equipment, such as a gauge roll whose length is also the width of the

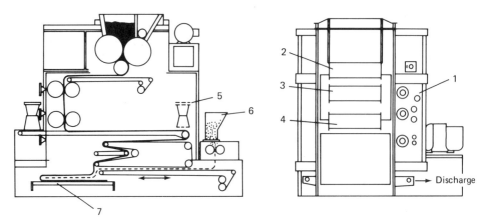

Fig. 8.10   Typical vertical laminator with continuous layering section (Courtesy of APV Baker, UK) (1) Facia panel, (2) Sheeter bottom roll, (3), (4) Gauge rolls, (5) Flour duster, (6) Fat/flour mixture duster, (7) Take-away carriage.

oven band, ranging from 500 mm to 1500 mm. The speed of the carriage web is about 30 to 45 mm per min.

The fat/flour mixture is free flowing to allow a duster to distribute it evenly onto the reciprocating carriage. The dough sheet is continuously folded on the slow-moving discharge conveyor, and the fat/flour mixture is sandwiched between layers. The compound movement of the discharging conveyor and the reciprocating carriage lays the continuous dough sheet backward and forward across the discharging conveyor in a zig-zag pattern of alternate 'triangles' on the upper and lower surfaces of the original sheet [4]. The zigzag folded sheet may be built up to about 10 to 12 layers [11]. It is then fed by the discharging conveyor to the second gauging line that precedes a biscuit cutter. The slower the discharge conveyor runs relative to the reciprocating carriage movement, the more layers will be formed in the final dough sheet.

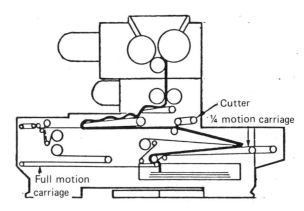

Fig. 8.11   Vertical laminator with cut sheet layering section (Courtesy of APV Baker, UK).

Some machines have two variable speed motors to drive the sheeter unit and continuous layering unit respectively via a transmission mechanism; others have additional motors such as a three-roll motor, a gauging roll motor, a transfer web motor, a relaxation conveyor motor, a main web motor, and a carriage drive motor. In any case, the sheeter, gauge rolls, and each web should be adjustable in a stepless speed mode to synchronize with each other as well as to ensure steady feeding at a rate required by the next gauging stage section.

Modern laminators are fitted with an advanced control system with a facia panel mounted on the front of the machine (Fig. 8.10) to allow the operator to regulate easily the sheeter speed, sheeter or gauge roll gap, speed variation for the web prior to the gauge rolls, speed variation for the layering web, and carriage speed, as well as to start/stop the machine normally or in an emergency. Table 8.5 gives specifications of this kind of machine.

The continuous layering process provides a continuous and smooth action to the dough sheet, and it requires a relatively simple construction of the layering unit. But the edges are always thicker than the middle part of the sheet since it is folded over at the sides. In this way, stress is introduced into the laminated dough. Furthermore, the fat is dusted only between every other lamination, accompanied by some exposure at the top surface of the zigzag pattern.

8.4.1.2   *Vertical laminators with a cut-sheet layering unit*

Cut-sheet layering can be performed in several different ways. The Vicars Group Ltd use a method which is described by D. J. R. Manley in *Technology of biscuits, crackers, and cookies* (1983; 2nd edition 1991). Here we shall describe the APV Baker system.

In Fig. 8.11 the sheeter and gauge rolls are similar in performance to those previously described. The cut-sheet layering unit comprises a rotary cutter to cut the

Table 8.5   Specifications of a typically vertical laminator with a continuous layer-ing section (APV Baker, UK)

	Sheeter			
			Bottom gauge	
Roll diameter	Forcing roll	Top gauge roll	roll	Gauge rolls
(mm)	420	420	230	230
Roll gap (mm)	5 to 15			0 to 10
Layering stroke	812 mm or 1016 mm			
Carriage web speed	36 m/min max.			
Carriage traverse speed	18 stroke/min max. at 1016 mm			
Main drive motor	7.5 kW			
Sheeter drive motor	7.5 kW			
Dimensions (mm)	2655 × 3595 × 2792			
Net weight (kg)	11300			

continuous dough sheet into lengths, a reciprocating full motion carriage which lays down the cut sheet, a reciprocating quarter motion carriage, and four driven webs.

The dough sheet is formed by the three-roll sheeter and the vertical gauge rolls (one or two or three sets) at the upper part of the laminator, and is rested and transferred to the main web, where it is cut to length by the layering stroke which is the width of the cutting machine on the biscuit production line. The reciprocating quarter motion carriage acts as a reservoir to receive the cut sheet, and then to send it back onto the main web before discharge. The main web is part of the full motion carriage. As the main carriage finishes its return stroke, the cut sheet is discharged onto the feeding conveyor of the next stage such as the second gauging roll section or the cutting machine. On the forward stroke, the sheet is carried over the take-away web as the reservoir web pays out the stored sheet. On the return stroke of the main carriage, the dough sheet is discharged onto the take-away web, while the reservoir web takes up a new supply of dough sheet ready for the next forward stroke.

The rotary cutter of the laminator is driven by the reservoir carriage drive to ensure synchronization. The position and the depth of cut can be adjusted by handwheels at the front of the machine via a phasing device and other mechanisms while the machine is running.

The quarter motion carriage drive also actuates the full motion carriage which incorporates an adjustable delivery nose roller to ensure that the laminated sheets are fed centrally to the take-away web which is also the feeding web for the next gauge rolls. The web is usually supported by a profiled feed table which can be adjusted at each corner to give the optimum setting while the machine is running.

The cut-sheet layering process provides smooth laminated edges without introducing stress as in the continuous layering mechanism. But the unsealed edges may result in fat leakage, and the cutting mechanism makes the machine a more complicated structure.

The machines so far described in this chapter have many similarities in construction and control. Lack of space precludes further discussion of vertical laminators, so we now pass on to another type of laminator.

8.4.2   A combination of MM laminator and SM stretcher

This combination was invented by Mr T. Hayashi, of Rheon Automatic Machinery Co. Ltd, in 1974. It differs from conventional laminators in two ways. Firstly, the bulk dough and the fat are extruded in the shape of a hollow tube rather than a dough sheet dusted with fat mixture. Secondly, the thickness of the laminated sheet is reduced by a continuous chain of stretching rollers rather than by pairs of rolls.

This machine is placed at the head of the production line for pizza, pie cookings, coffee cakes, butter rolls, doughnuts, etc. Currently, it is used to work on dough for high-grade biscuits by stretching out the cracker type puff pastry into 1 mm thickness which is then cut off and baked. With this machine, alternate layers of fat and dough of uniform thickness are stretched, layered again, and then restretched, the layering being repeated many times, until the desired number is obtained with a double sheet of 4 mm thickness containing 96 layers of fat, in which the thicknesses of fat and dough layers are 13 μm and 29 μm respectively. The machine makes more elaborate products

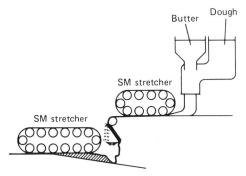

Fig. 8.12   Laminating and stretching process (Courtesy of Rheon Automatic Machinery Co. Ltd).

more accurately than they can be made by hand. It can pile up hundreds of layers of butter and dough in which the thickness is as thin as 5 μm per layer and can roll the layers into a very thin sheet by means of fully automatic units. Fig. 8.12 shows the laminating and stretching process which is carried out by a laminator, pile-up table, and stretchers.

8.4.2.1   MM laminator

This unit comprises two hoppers, one for dough and the other for fat, a compound nozzle, a small stretcher for first forming, and a folding mechanism for continuous layering.

The compound nozzle is a small cylinder for forming the dough and fat into a hollow tube, placed at the head of the system (Fig. 8.13), and it prepares the layers of

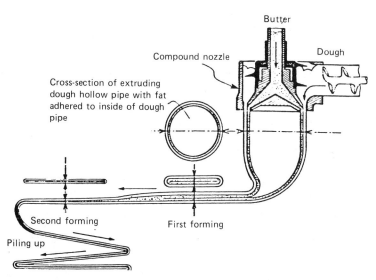

Fig. 8.13   Compound nozzle, hollow tube, and sheet sandwiched with fat (Courtesy of Rheon Automatic Machinery Co. Ltd).

fat and dough. As the dough goes into the horizontal chamber from a separate hopper, it is conveyed and extruded by the dough screw out of the outer ring of the compound nozzle in a tube-like shape, while the fat is simultaneously fed from another hopper uniformly upon the inner wall of the hollow dough tube.

The small stretcher spreads the tube sideways, forming a continuous layer of fat and dough of even thickness. This is the first forming stage.

After the thickness of the dough layers sandwiched with fat have been adjusted to 3 or 4 mm, the dough is continuously folded by the folding mechanism to pile up to a height ranging from 40 to 150 mm before proceeding to the next stretching stage. Table 8.6 shows specifications of MM laminators.

8.4.2.2   SM stretcher

This is the most significant invention in the Rheon MM production line. It consists of a special rolling mechanism composed of a group of rollers turning on their axes and revolving around the circumference of the group at the same time (Fig. 8.14).

As the folded dough is engaged with the second stretcher, it is stretched, in one operation, into a thin unbroken sheet ranging from 3 to 30 mm in thickness. If more layers are required, the dough may be folded and stretched again to reach the desired number of layers, and the thickness of the sheet may be reduced to 1.5 mm.

8.4.2.3   Sheeting principle of the SM stretcher

It is necessary to look at the conventional roll system before discussing the SM stretcher sheeting principle. As the piled dough is rolled by means of the conventional system, so-called multi-laminar vortexes appear as shown in Fig. 8.15. Before the dough enters the nip between the rollers (bb′), the lines of force (l) form large vortexes. In this conventional roll system, friction is required on the roll surface to compress the dough into a continuous sheet. The larger the thickness reduction, the more severe is the compression necessary on the dough sheet, which may exceed 20 kg/cm² and may result in an increase in the dough speed of flow which will exceed the roll surface speed. In this case, either violent vortexes or a concentration of pressure may occur, leading to irregular folds and distortions. So the conventional roll system (gauge rolls) is admirable for the preparation of crisp biscuit dough sheeting, but less good as the SM stretcher for precise sheeting with a large thickness reduction.

Table 8.6   Specifications of MM laminator (Rheon Automatic Machinery Company. Ltd)

Model	MM-1	MM-2	MM-3
Output (kg/h)	230–350	300–500	500–1000
Electric power (kW)	3.135	3.135	4.635
Number of folds	16 layers at max.		
Dimensions (mm)	3035 × 1230 × 2030		

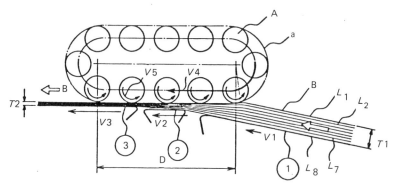

Fig. 8.14   Sheeting principle of SM stretcher (Courtesy of Rheon Automatic Machinery Co. Ltd.)

As shown in Fig. 8.14, the folded dough sheet travels across the three conveyors (1), (2), and (3) of the SM stretcher, moving at different speeds V_1, V_2, and V_3. It is subjected to continuous tensile stress of about 70 g/cm² which does not harm the dough. The stretcher rollers turn on their axes and revolve around the circumference of the entire group at the same time, so that the accentuated pressure with little friction is applied additionally from above on the dough underneath. As a result, the dough is continuously stretched and reduced from thickness T_1 to a preset thickness T_2, which is expressed by V_1/V_2.

For the conventional roll system, the stress produced at the narrow portion between the rolls can exceed 20 kg/cm² to reduce the thickness of the dough sheet, while in the SM stretcher it is only about 70 g/cm². 'The reason for this is that in the case of the SM stretcher the tensile stress is well balanced with the stress resulting from the pressure above." (Rheon Automatic Machinery Co. Ltd.)

The MM laminator, pile-up table, and SM stretcher are part of the Rheon MM line system. The following principal equipment consists of a piler, make-up table (MT), safety devices, and control. This MM line is widely used throughout the World for efficient production of various kinds of bread, high-grade biscuits, etc.

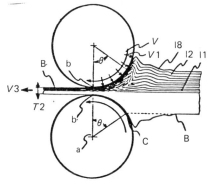

Fig. 8.15   Large thickness reduced by conventional roll system (Courtesy of Rheon Automatic Machinery Co. Ltd).

9

Reciprocating cutters

Traditional reciprocating cutters are still the dominant type for biscuit cutting operations in China, even though more than twenty years have passed since the first rotary cutter was introduced.

9.1 INTRODUCTION

The early type of reciprocating cutter worked in a vertically reciprocating mode by which the dough sheet was cut into pieces as it was moved forward step-by-step by a set of pawl and ratchet wheel mechanisms. This type of cutting machine usually works at a low speed, with a low machine output, since a higher speed will cause inertia and stronger shocks and will result in uneven thickness along the sheet width and some cracks along the sheet sides. Furthermore, this type of machine is not suitable for a continuous biscuit production line, especially as it cannot match the continuous channel oven band.

To overcome these shortcomings, the early interrupted reciprocating cutter has been developed into a swinging reciprocating one by which the dough sheet is cut into pieces as it is travelling forward at a steady speed toward the oven, while the crosshead is swinging forward with the dough sheet conveyor to perform the cutting operation and then swings back to begin the next cutting action. Because of this the machine is often referred to as a swing-reciprocating cutter or swing cutter for short. Machines of this type have been employed in the biscuit industry for several decades, and are not likely to be replaced completely in the near future, owing to their low cost, ease of manufacture, convenience of maintenance, and adaptability for a wide range of products by simple alteration of the crosshead.

9.2 CONSTRUCTION

The swing reciprocating cutter comprises a dough piece shaping assembly, drive and transmission mechanisms, frame, and control systems.

Fig. 9.6 shows a typical reciprocating cutter placed in the production line between the dough sheet relaxation and the scrap pick-up conveyors.

9.2.1   Crosshead

This unit is part of the dough piece shaping assembly, which generally consists of one, two, or more rows of individual cutters including printers and knives mounted on the cutter support (4) across the width of the cutting web underneath (Fig. 9.1). The bottom face of the crosshead is fixed with a flat metal stripper plate (5) to assist separation after cutting.

A typical cutter and printer-knife assembly is designed as shown in Fig. 9.2. The printer is usually made of brass mounted on the support plate (5) by means of connecting plate (9), bolt (6), and nut (1) via the upper connecting plate (3), spring (4), spacing collar (7), and screws. The knife (8) is also fixed on the support plate (5) by screws, and the edges of the knife are shaped like the outer ring of the printer.

Most printers are designed with dockers to make holes that go right through or half way through the biscuit thickness. The holes allow steam to escape from the underside of the cut dough piece during baking to prevent a dome-like top surface and sunken bottom. The docker holes also assist heat to penetrate more evenly, which results in uniformly baked biscuits with uniform thickness to allow them to be mechanically packed to a specified height.

The dockers can be either integral on the printer or are made from thin brass rods inserted into the printer after graving. The latter method makes manufacture and repair cheaper and easier.

Printers come in two main forms, light and heavy.

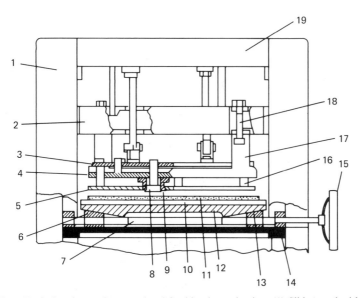

Fig. 9.1   Typical reciprocating crosshead for biscuit production. (1) Slide stand with slide groove, (2) Slide bridge, (3) Upper connection plate, (4) Cutter support, (5) Stripper plate, (6) Left adjustment device, (7) Double-ended screw, (8) Printer, (9) Knife, (10) Cutting web, (11) Dough sheet, (12) Pad table, (13) Right adjustment device, (14) Support, (15) Handwheels, (16) Connecting bolt, (17) Crosshead assembly, (18) Connecting bolt, (19) Upper bridge.

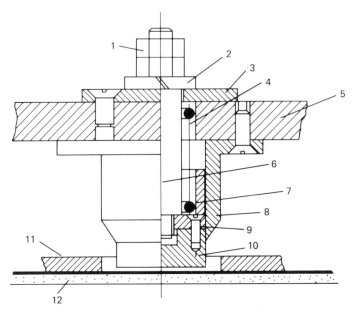

Fig. 9.2   Single printer-knife unit [44] (1) Nut, (2) Elastic ring, (3) Upper connecting plate, (4) Spring, (5) Crosshead support, (6) Connecting bolt, (7) Spacing collar, (8) Knife, (9) Connecting plate, (10) Printer, (11) Stripper plate, (12) Dough sheet.

The dough for crackers is more elastic than that for semi-sweet biscuits, which makes it difficult to keep a complicated pattern but allows gas expansion so that the printer for crackers is usually designed with only simple relief patterns such as letters and with more docker pins. This kind of printer is often referred to as a 'light printer' with a weaker spring force (Fig. 9.2 (4)) by which the biscuits always have light sunken patterns and more docker holes.

The dough for semi-sweet or some sweet biscuits is more plastic than that for crackers, which allows more complex and pleasing patterns to survive baking, so that the printer for semi-sweet or sweet biscuits is generally designed with more complicated gravure patterns and fewer docker pins with a greater spring force (Fig. 9.2 (4)). This kind of printer is often referred to as a 'heavy printer' by which the biscuits always have heavier raised patterns and fewer docker holes. The number of docker pins is also dependent upon the size of the product and the type of oven band to be used such as a steel band, perforated band, wire mesh, or heavy mesh band.

The usual width of the crosshead, designed in conjunction with band ovens, is from 400 to 1250 mm. The cutters can be arranged in one, two, three, or even four rows, mainly depending on the size of a single cutter unit. To some extent, the more rows of cutters that are mounted on the crosshead, the greater the output would be without increasing the stroke speed of the crosshead. However, too many rows of cutters sometimes result in uneven cutting pieces, increasing crosshead weight and power consumption for each stroke. The space on each side of the crosshead should be about 20 to 25 mm to allow continuous trims on both sides of the dough sheet, which is convenient for the separation and scrap return after the cutting operation.

There are two types of layout for the arrangement of cutter units on the crosshead across the width of the cutting web. One is a lateral staggered layout, and the other a checkerboard layout. For square and rectangular products both layout methods are available, while for round or oval products only the former pattern is suitable. Otherwise, too much diamond-like clear space would be left at the centre of each square of four pieces, leading to an ineffective use of the dough sheet and perhaps to overheating of the biscuit edges at that point [4].

The gap between cutters on the cutter block is also dependent on the behaviour of the dough piece in the oven. Scrapless cutting or very close spacing on the block is used for doughs which shrink during baking, while the looser spacing is used for doughs which expand in the oven.

9.2.2 Cutting web and pad table

The cutting web under the crosshead is also very important for the dough piece shaping. It is usually made of heavy canvas or food quality synthetic materials. The web should be fairly tight above the pad table to reduce distortion of the cut pieces. The cutting web is often used wet, to provide a sticky surface to assist the separation of the dough piece and the printer after cutting by means of an oiling device (see Fig. 10.3 in the next chapter).

The pad table, usually made of hardwood, is placed under the cutting web to act as a supporting table for cutting and printing operations. The characteristics of the table are very important for precision and high quality shaping functions. As Dr Samuel A. Matz pointed out: 'The more the pad yields to cutter pressure, the more the dough is pushed up into the centre of the die (printer). It tends to cling more to the edges of the cutter and the docker pins. Border designs and brand names or symbols are impressed too deeply. Release is interfered with and the dough sheet is distorted into folds and ridges' [8]. That is why the pad table is usually made of hardwood, sometimes with cross-grain presented to the printer (Fig. 9.1 (12)).

9.3 Shaping principle

Driven by the swing and sliding mechanisms, the crosshead performs the printing and cutting operations in its swinging down stroke as the dough sheet moves forward on the cutting web at a steady speed.

The stripper plate, mounted on the bottom face of the crosshead assembly, stops just above the top surface of the dough sheet as the crosshead descends (Fig. 9.2). But the printer goes further to make patterns on the top surface of the dough sheet, and the inside spring (4) is compressed to buffer the stamping forces. Since the depth of the pattern either for a sunken pattern or a raised one is very limited, the printer has a small movement, while the cutting knife has to go through the whole depth of the dough sheet to cut it into pieces. By that time, the working stroke has been completed, and the cutting depth (the thickness of the sheet) is the cutter effective travel L_0 (Figs 9.3, 9.4, 9.5).

During the up stroke, the knife and the printer move upward to leave the dough piece and strip behind as the stripper plate remains above the sheet. Separation of the printer is easily carried out, since some release agent has been employed. The spring (4) (Fig. 9.2) is released during the up stroke; it produces a pushing force to make the dough piece stick onto the wet cutting web. Separation of the cutting knife, however, is less easy, and sometimes it causes trouble as the newly cut circle surface tends to stick on the metal surface as the knife leaves. The stripper plate is designed to assist the separation by 'stripping' the dough scraps from the knives and allowing them to fall back onto the cutting web as the knife and the printer rise through the plate. Hence the name 'stripper plate'. After the dough is stripped off, the stripper plate is raised and swings forward at the same time with the crosshead ready for the next cutting stroke, while the cutting web carries the dough pieces steadily forward to the panning unit. Thus, as the crosshead begins its next working stroke, a clear space of dough sheet is prepared under the crosshead for the successive shaping process.

There are four sets of springs inside the upper bridge (19) (Fig. 9.1) to buffer the shocks of the shaping operation.

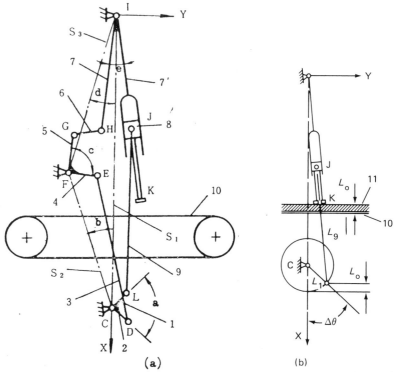

Fig. 9.3   Crosshead drive; action, X-Y coordinates [15] (1) Cutting crank, (2) Swing crank, (3), (6), (9) Connecting rods, (4), (5), (7), (7') Swing levers, (8) Crosshead slide, (10) Dough sheet (piece) conveyor, (11) Dough sheet.
S_1, S_2, S_3 Frame lengths of the lever system.
a, b, c, d, e Installation angles between levers.
C, D, E, F, G, H, I, J, L Articulated joints of levers K Cutter,
L_0 Effective travel, dough piece thickness.
L_1 to L_9 Lever and crank lengths.
$\Delta\theta$ Required rotation angle.

9.4   Crosshead drive

Fig. 9.3(a) shows the crosshead drive in an X-Y coordinate system, and Fig. 9.3(b) is a schematic diagram of the crosshead cutting operation.

The crosshead drive unit consists of a slider crank chain, a double-rock lever mechanism, and a crank-swing-slide system.

As the machine is switched on, the spindle, via the articulated joint (C) (Fig. 9.3(a)), actuates both the swing crank (2) and crank (1) to rotate at the same time. By means of the correcting rod (3), swing rods (4) and (5), crank (2) drives the crosshead swing rod (7) to swing reciprocally forward and backward as crank (1) makes the crosshead slider (8) move in a linear reciprocating mode along the guide groove of swing rod (7) via connecting rod (9). Therefore the cutter K (Fig. 9.3(b)) on the crosshead can move forward with the cutting web and slide down to cut the sheet into pieces and then goes up as it swings back for the next cutting operation. The cutting web, driven by a variable speed motor via a very simple transmission system, moves forward at a steady speed so that a clear space of dough sheet is left under the crosshead, which allows the next shaping operation to begin.

The precision action of the crosshead swing movement and the synchronization of the crosshead with the cutting web are very important for biscuit appearance and quality and for packing feasibility, especially for thicker biscuit production. A poor design will result in an unclear pattern and untidy edges to the cut dough pieces. Fig. 9.4 shows a swing cutter movement oval orbit (a) and the alteration of its horizontal velocity component (b), which is quite typical but not good enough [12].

In Fig. 9.4(a), the major axis of the ellipse is at 35° to the horizontal cutting web, which makes the cutter down stroke larger than the cutter up stroke. That is, the travel from the beginning of contact with the dough sheet point (1) to the lowest point (2) (the cutter descends further down through the dough sheet but stops just at the top surface of the cutting web without cutting it) is longer than the cutter up stroke from the lowest point (2) to point (3) at which the cutter leaves the dough piece. The unbalanced stroke causes uneven contractions in the shaping process.

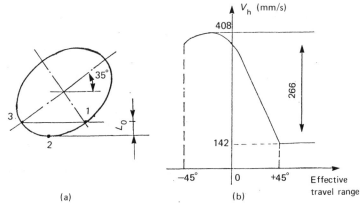

Fig. 9.4(a)   Typical cutter movement orbit, (b) the horizontal velocity component.

In Fig. 9.4(b), the horizontal velocity component V_h of the cutter top point is changing in the range from 408 mm/s to 142 mm/s, which is unsymmetrically distributed, while the sheet is conveyed in a uniform linear motion by the cutting web at a speed of 213 mm/s. Under these conditions it is difficult to make dough pieces with clear patterns and tidy configurations at a higher speed, because of the large differences in the velocities of the cutter and the cutting web as well as the unbalanced up-and-down strokes of the cutter. Therefore, some work has been done to optimise the design [12].

As shown in Fig. 9.3, the crosshead driving action is carried out by the four-lever mechanisms CDEF and FGHI as well as the five-lever mechanism IJLC. If the lengths of the cranks and levers from 1 to 9 are L_1 to L_9 respectively, and the maximum thickness of the dough sheet is the cutter effective travel L_0 (Fig. 9.3(b)) from contact with the dough to cutting to the maximum depth, then the required rotating angle $\Delta\theta$ of crank (1) for cutting the sheet into pieces can be approximately calculated from equation (9.1).

$$\Delta\theta = \cos^{-1}\frac{L_1 - L_0}{L_1} \tag{9.1}$$

The design objective for the swing reciprocating drive is to achieve the minimum difference between the speeds of the cutter and the cutting web. For the crosshead drive system, the object of optimizing is to determine the lengths of the cranks, levers, and frames as well as the installation angles of the elements.

Fig. 9.5 shows the cutter movement orbit (2) and its horizontal velocity component curve (b) after computer optimization of the previously discussed reciprocating cutter.

In Fig. 9.5(a), the major axis of the ellipse of the cutter movement orbit is vertical to the horizontal cutting web and symmetrically distributed. In Fig. 9.5(b), the horizontal velocity component of the point K on the cutter is changing gently in the range from 167 to 240 mm/s. Compared to Fig. 9.4, the velocity alteration is reduced by 73%, so that the cutter movement is more in synchronization with the dough sheet (cutting web) movement. With the optimized system, high quality products, with clear patterns and tidy edges, can be made at high speed.

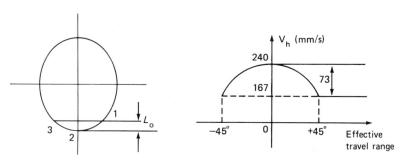

Fig. 9.5   Cutter movement orbit and its velocity component curve after optimization [12].

9.5 TRANSMISSION

In China, most reciprocating cutters are provided with gauge rolls driven by a common variable speed electric motor via belt and pulleys, chain drives, and gear trains. Fig. 9.6 shows a typical reciprocating cutter with three sets of gauge rolls. Fig. 9.7 is a schematic transmission diagram of this type of biscuit cutting machine. Scheme 9.1 is the transmission expression.

$$\text{Motor 1} \rightarrow \text{Shaft I} \rightarrow \frac{\text{Pulley 2}}{\text{Pulley 3}} \rightarrow \frac{\text{Gear 4}}{\text{Gear 5}} \rightarrow \text{Shaft II}$$

Shaft II
$$\begin{cases}
\text{Eccentric wheels 8,9,10, and 11} \rightarrow \text{Crosshead up and down and swinging} \\[2mm]
\dfrac{\text{Gear 6}}{\text{Gear 7}} \rightarrow \text{Shaft III} \rightarrow \dfrac{\text{Chain wheel 17}}{\text{Chain wheel 18}} \rightarrow \text{Shaft IV}
\end{cases}$$

Shaft IV
$$\begin{cases}
\text{Cutting web roll 23} \\[2mm]
\dfrac{\text{Gear 19}}{\text{Gear 20}} \rightarrow \text{Shaft V} \rightarrow
\begin{cases}
\text{Bottom roll 24 of first pair of gauge rolls} \\[2mm]
\dfrac{\text{Gear 21}}{\text{Gear 22}} \rightarrow \text{Top roll 25 of third pair of gauge rolls} \\[2mm]
\dfrac{\text{Chain wheel 26}}{\text{Chain wheel 27}} \rightarrow \text{Shaft VII}
\end{cases}
\end{cases}$$

Shaft VII
$$\begin{cases}
\text{Bottom roll of second pair of gauge rolls, 32} \\[2mm]
\dfrac{\text{Gear 29}}{\text{Gear 30}} \rightarrow \text{Shaft VIII} \rightarrow \text{Top roll 33 of second pair of gauge rolls} \\[2mm]
\dfrac{\text{Chain wheel 31}}{\text{Chain wheel 34}} \rightarrow \text{Shaft IX}
\begin{cases}
\text{Bottom roll 37 of first pair of gauge rolls} \\[2mm]
\dfrac{\text{Gear 35}}{\text{Gear 36}} \rightarrow \text{Shaft X} \rightarrow \text{Top roll 38 of} \\ \qquad\qquad\qquad\qquad\qquad \text{third pair of} \\ \qquad\qquad\qquad\qquad\qquad \text{gauge rolls}
\end{cases} \\[2mm]
\dfrac{\text{Chain wheel 28}}{\text{Chain wheel 39}} \rightarrow \text{Shaft XI} \rightarrow \text{Scrap pick-up web roll 40}
\end{cases}$$

Scheme 9.1.   Transmission scheme for a biscuit cutting machine.

From motor (1), the rotation is transmitted through pulleys (2) and (3), gear wheels (4), (5), (6), and (7) to shaft II.

The eccentric wheels (cranks) (8), (9), (10), and (11) are mounted on shaft II to actuate the crosshead to swing and to travel up and down for cutting the dough sheet into pieces.

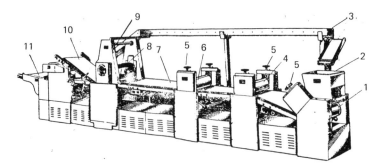

Fig. 9.6 Typical reciprocating cutter with three sets of gauge rolls for biscuit production (Courtesy of Tal Zhou Food Machinery Factory, China) (1) First gauge rolls, (2) Dough hopper, (3) Scrap side return, (4) Second gauge rolls, (5) Handwheel for gap adjusting, (6) Third gauge rolls, (7) Dough sheet relaxation conveyor, (8) Slide bridge of reciprocating crosshead, (9) Upper bridge of crosshead, (10) Scrap pick-up conveyor, (11) Oven feed conveyor.

From shaft III, the rotation is transmitted to cutting web roll (23) and the third pair of gauge rolls (24) and (25) via chain wheels (17) and (18), gear wheels (19), (20), (21), and (22) as well as shafts IV, V and VI.

Shaft V transfers the rotation through chain wheels (26) and (27), gear wheels (29) and (30) as well as shafts VII and VIII to the second pair of gauge rolls (32) and (33).

On both ends of shaft VII are mounted chain wheels (28) and (31) by which the rotation is transmitted via chain wheels (34) and (39) as well as gear wheels (35) and (36) to the first pair of gauge rolls (37) and (38) and to the scrap pick-up web roll (40) from which the scrap is transferred to the side return web leading to the gauge roll hopper for reuse.

9.6 Adjustment and capacity

For optimum printing and cutting, the printer travel and the height of the cutting web should be adjusted properly.

The printer travel is controlled by changing the spacers of the spacing collar (7) in Fig. 9.2. The cutter level is fixed before operation, but the dough sheet level, which is the height of the cutting web, is regulated during operation by means of handwheel (15) (Fig. 9.1) via the left and right adjusting wedge devices (6) and (13) and the screw (7). This screw is threaded on both ends, one is left-handed and the other right-handed, which allows the left and right adjusting devices to raise or lower, respectively, the cutting web height (that is, the level of the dough sheet).

The capacity of a reciprocating cutter can be calculated according to equation (9.2)

$$G = 60 \, N_s \, P \, \eta_s / K \qquad (9.2)$$

Where G is the capacity of the machine, kg/h; N_s is the number of strokes per minute; P is the number of dough pieces per stroke; K is the number of biscuits per kilogram after baking; and η_s is the standard rate of biscuits suitable for packing, which is less

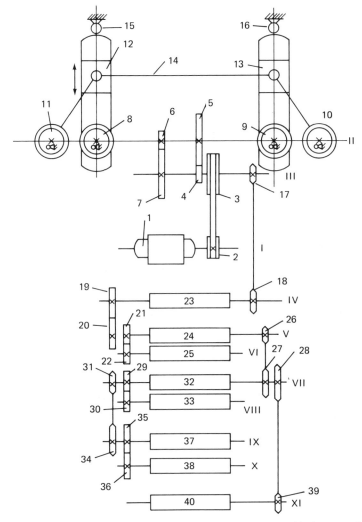

Fig. 9.7 Typical transmission system of a small reciprocating cutter with three sets of gauge rolls (Courtesy of Xin Xiang County Food Machinery Factory, China) (1) Variable speed motor, (2), (3) Pulleys, (4), (5), (6), (7) Gear wheels, (8), (9), (10), (11) Eccentric (crank) wheels, (12), (13) Crosshead slides, (14) Slide bridge, (15), (16), Articulated joints, (17), (18) Chain wheels, (19), (20), (21), (22) Gear wheels, (23) Cutting web roll, (24), (25) Third pair of gauge rolls, (26), (27), (28) Chain wheels, (29), (30) Gear wheels, (31) Chain wheel, (32), (33) Second pair of gauge rolls, (34) Chain wheel, (35), (36) Gear wheels, (37), (38) First pair of gauge rolls, (39) Chain wheel, (40) Scrap pick-up web roll. I–XI Shafts.

than 100% because some biscuits may be overbaked, domed, or cracked and not suitable for marketing.

No doubt the swing reciprocating cutter has made great progress from the old vertical reciprocating model. It gives smooth and steady cutting and printing operations. However, too high a cutting rate can cause vibration and noise, so that a

Table 9.1   Specifications of some small reciprocating cutter units

Model	1	2	3
Width of band	560	480	400
Capacity (kg/h)	up to 300	up to 240	200
Strokes per min.	120	120	120
Variable speed motor	2.2 kW	2.2 kW	1.1 kW
Dimensions (mm)	2885 × 1000 × 1535	3370 × 820 × 1708	2320 × 770 × 1210
Net weight (kg)	1800	1240	1000

strong floorbase and foundation bolts are often needed to buffer the possible vibration and to reduce noise. For the same reasons, the stroke rate is commonly under 180 strokes per minute in China.

As shown in Fig. 9.2, the cutter and the printer and knife are separate and are assembled together. Thus, if any damage occurs to a docker or a printer or a cutting knife, it is easier for them to be repaired or replaced than for a rotary cutter.

Generally speaking, a set of crossheads is much cheaper than a pair of rotary cutter rolls. The reciprocating cutter allows easy changing of the crosshead or individual printer-knife units, so that the unit can be readily adapted for different products.

In China, newly baked goods are welcomed by consumers. In the rural areas, food factories prefer small-scale machines to produce colourful biscuits, so that many food machinery factories now develop their machines not to a greater size but on a smaller scale to meet the needs of the large rural market. Table 9.1 shows the general specifications of some small swing reciprocating cutters.

10

Rotary cutters

10.1 INTRODUCTION

The rotary cutter was invented in the 1970s. It completely changed the biscuit cutting method from the traditional reciprocating action into a rotary mode. Most of the difficulties found with the reciprocating cutters have been overcome by this simple alteration.

The early rotary cutter comprised a shaping roll and an anvil roll. The shaping roll, with many cutter shapes and sharp edges, performed the printing and cutting actions simultaneously, as the dough sheet travelled forward and passed between the top shaping roll and the bottom anvil roll. There was no device such as a stripper plate or scraper knife to assist separation of the dough and the shaping roll. Also, the deformation area of the patterns and the newly cut surfaces have a tendency to stick on the cutter roll surface, especially when more pressure was needed to carry out the shaping and cutting processes simultaneously. Adherence of the dough to the shaping roll often caused trouble. This problem was overcome by the second generation of rotary cutters—a three-roll cutter unit, in which the shaping process was separated into two stages performed by two separate printing and cutting rolls, placed above a common anvil roll. Since the common bottom roll provides the two top rolls with working areas, the bottom roll diameter is much greater than that of the upper rolls. The greater the diameter of the roll, the bigger the machine will be, and similarly with the machine tools needed for manufacture and the maintenance tools.

The third generation of rotary cutter is the four-roll rotary cutter unit which comprises a pair of driven rolls, consisting of a printing and a cutting roll with their respective rubber covered anvil rolls. In the production line that was shown in Fig. 8.4 (6) is a typical rotary cutting machine with two top rolls and cutting web with one or two anvil rolls.

10.2 CONSTRUCTION

The rotary cutting machine is mechanically very simple compared with the reciprocating cutter unit. It generally comprises shaping rolls, bottom roll(s), and a simple transmission system driven by a variable speed motor rather than by complex lever mechanisms as in the reciprocating cutter unit.

Modern machines provide an advanced control system which allows remote servo adjustment with electronic measurement and digital readout suitable for computer control systems. They are also equipped with a facia panel mounted on the side of the machine. There are buttons on the facia for the operator to start or stop the machine, to adjust the cutting and printing pressure, to regulate speed, and to phase the rolls.

10.2.1   Printing and cutting rolls

These two rolls are usually housed in a cradle frame which in turn is supported in the main frame. The roll diameter ranges from 105 to 305 mm (200 to 300 mm in China), mainly depending upon the size of the biscuit and the machine capacity required. The roll length is designed to suit the width of the oven band, ranging from 400 to 1250 mm (or more).

Fig. 10.1 shows (a) a printing and (b) a cutting roll, made of cast bronze, respectively engraved with the required pattern and with the appropriate cutting edges. The rolls have steel stub shafts which are mounted on frames via sealed ball bearings and are driven via gear wheels keyed to the end of the roll shaft.

The printing and cutting stages are synchronized by a phasing mechanism which ensures that the cut piece exactly matches the printed design.

The manufacture of the impressions on the metal tube is complicated, expensive, and time-consuming. The impressions are sometimes coated with non-stick materials to assist the separation of the dough and the impressions. However, the coatings are not everlasting. They may wear off in a few months.

The edges on the cutting roll should be kept sharp to give a clean cutting action, since it is done during rolling rather than by a swing reciprocating action. Furthermore, the machine is often required to provide smooth and steady performance at a much higher speed than the reciprocating cutter. Therefore both the printing and cutting rolls need expensive maintenance or replacement.

These problems are mitigated by the application of plastic moulds, which are easier to make and cheaper to replace. Fig. 10.2 shows plastic moulds which are screwed to

(a)

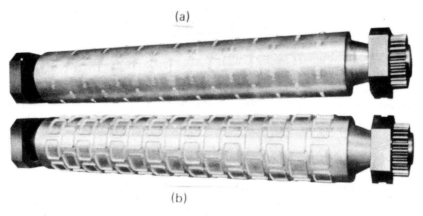

(b)

Fig. 10.1   Typical printing (a) and cutting (b) rolls (Courtesy of APV Baker, UK).

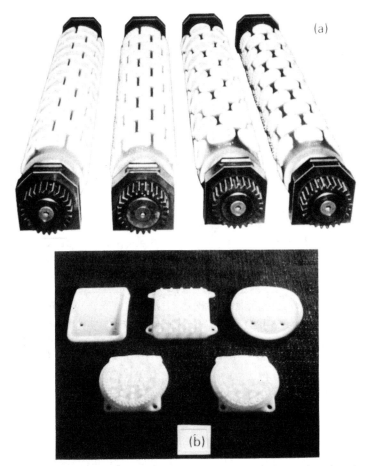

Fig. 10.2   (a) Printing and cutting rolls with raised plastic impressions and cutting edges. (b)
Individual plastic moulds (Courtesy of APV Baker, UK).

the periphery of a steel tube. The tube is fitted with steel stubs at both ends which are supported and driven in the same way as the metal rolls.

The number of impressions on the roll is dependent on the roll dimensions, the size of the biscuit design, and on the dough properties that are related to oven behaviour, which allow either scrapless cutting or very close or loose spacing.

10.2.2   Anvil roll and cutting web

As shown in Fig. 10.3, the anvil roll is used to support the moving dough sheet via the cutting web as the top rolls perform the printing and cutting operations. The steel anvil roll is covered with thick food quality rubber (10 to 20 mm) which plays the roll of an elastic pad table and helps to 'suck' the dough pieces from the cutters.

The characteristics of the cutting web and the rubber covered roll are very important for successful cutting, since too soft a pad will make dough pieces irregular

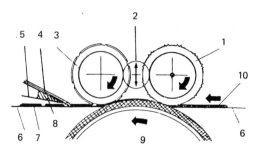

Fig. 10.3   Shaping principle of a three-roll cutter unit with a single cutting web without a 'band break'. (1) Printing roll, (2) Phasing mechanism, (3) Cutting roll, (4) Scraps, (5) Scrap pick-up web, (6) Cutting web, (7) Cut dough piece, (8) Nose piece, (9) Rubber covered anvil roll, (10) Dough sheet.

and deformed, while too dry a cutting web will not pull the cut dough pieces off the cutting roll, so that they will be carried up and over the roll. Hence, the rubber should be hard enough and the cutting web should be kept wet enough to provide a sticky surface for the dough to adhere to. Fig. 10.4 shows an oiling device used for wetting the cutting web. It consists of an oil container fixed on the sideframes by means of hanging supports, two oiling rolls driven by gears and a control mechanism.

The device is placed under the canvas cutting web before the shaping rolls. The oil feed roll (6) is half-submerged in the edible oil container. As gear (7) is engaged with

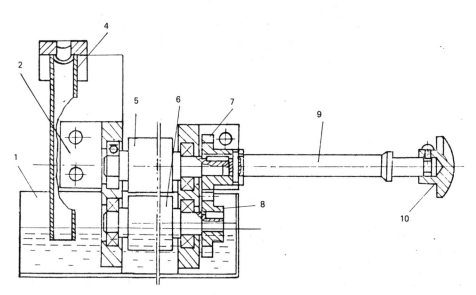

Fig. 10.4   Oiling device for cutting web (Courtesy of Xin Xiang Country Food Machinery Factory, China). (1) Edible oil container, (2) Hanging support, (4) Oil filling tube, (5) Web-oiling roll, (6) Oil feed roll, (7), (8) Gears, (9) Driving and adjusting shaft, (10) Control lever.

gear (8) by pushing down the control lever (10), the rotation will be transmitted to roll (6) via the adjusting shaft (9) which is driven by the other gearing. Rolls (5) and (6) rotate face to face, so that oil is fed to the cutting web by roll (5). The amount of oil fed to the web is regulated by the roll rotating speed. The faster the rolls revolve, the more oil is fed. To stop feeding, control lever (10) is pulled out.

The diameter of the anvil roll is about 200 to 250 mm for a four-roll rotary cutter unit and 400 to 450 mm for a three-roll rotary cutter unit.

10.3   Shaping principle

Fig. 10.3 showed the principle of a three-roll rotary cutter unit. The dough sheet is carried on the cutting web through the revolving printing and cutting rolls, and the web is supported on a single large diameter rubber covered drum. For this unit, the dough piece is shaped in two stages. Firstly, the top surface of the sheet is printed with the biscuit design by the top revolving printing roll as it travels and comes into contact with the printing roll. Secondly, the printed sheet is soon engaged with the revolving cutting roll and is cut into individual pieces with the printed design correctly placed, since the unit is equipped with a phasing mechanism which allows the operator to align the cutting edges with the printed patterns by the phasing adjustment wheel on the facia panel while the machine is running.

The anvil roll is driven with the moving web by friction. The position of the anvil roll can be adjusted in two directions to allow the printing and cutting pressure to be applied independently. The roll is first adjusted to impress the dough onto the printing roll until a satisfactory impression is achieved. It is then adjusted to bring the printed dough sheet into contact with the top cutting roll to obtain cleanly cut pieces.

Modern machines are also equipped with a pneumatically operated fast release mechanism which rapidly lowers the anvil roll simultaneously from both the printing and cutting rolls. This allows any inferior quality dough to pass through without causing any pick-up on the rolls and then returns the anvil roll to its former working position to continue normal operation.

In the four-roll rotary cutter unit, there are two bottom rubber covered rolls (Fig. 10.5) instead of one, with a pair of top printing and cutting rolls. The anvil rolls are often supported by pneumatic cylinders by which the height of the anvil rolls can also be adjusted separately. If any irregularities appear in the dough sheet, the anvil rolls can be lowered and lifted quickly by actuating hand operated valves without stopping the machine.

A phasing mechanism is also fitted on this unit with which the operator can align the cutting edges with the printed biscuit design by means of a handwheel.

The height of the top rolls is very important for precise shaping since too low a position will cause too great pressure on the dough, which would result in spread edges of the dough in the printing stage and a cut edge pressing into the anvil roll in the cutting stage. Too high a position would give an unclear impression of the design and unclean or even connected cut edges of dough. Therefore rotary cutting machines usually have adjustment facilities for the printing and cutting roll, operated by handwheels.

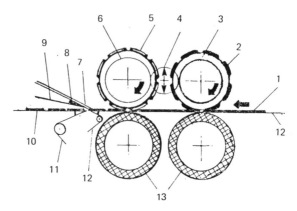

Fig. 10.5 Shaping principle of a four-roll rotary cutter unit. (1) Dough sheet, (2) Plastic impression, (3) Printing roll, (4) Phasing mechanism, (5) Plastic cutting edges, (6) Cutting roll, (7) Dough scraps, (8) Web nose-piece, (9) Dough scrap pick-up, (10) Cut dough piece, (11) Transfer web, (12) Cutting web, (13) Rubber covered anvil rolls.

10.4 TRANSMISSION AND CAPACITY

The rotary cutter unit is mechanically very simple. Its printing and cutting rolls are driven by a variable speed motor through a few gear wheels or chains with an incorporated mechanical reducer so that the bottom rubber covered rolls are driven by the cutting web by friction.

Fig. 10.6 shows a typical rotary cutter transmission system in which the variable speed motor transmits rotation and torque through a worm gearing reducer, belt and

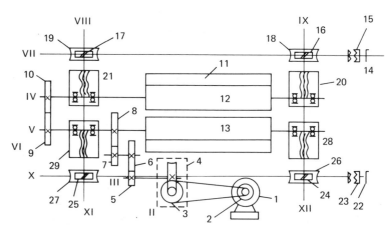

Fig. 10.6 Transmission system of a typical rotary cutter unit. (1) Variable speed motor, (2), (3) Pulleys, (4) Reduction box, (5)–(10) Gear wheels, (11) Anvil roll, (12) Printing roll, (13) Cutting roll, (14) Handwheel, (15) Clutch, (16), (17) Worms, (18), (19) Worm gears, (20), (21) Screw blocks, (22) Handwheel, (23) Clutch, (24), (25) Worms, (26), (27) Worm gears, (28), (29) Screw blocks, I–XII Shafts.

pulley, gear trains, and intermediate shafts to the printing and cutting rolls. These two top rolls revolve in the same direction and at the same speed so that the printing and cutting operations can be synchronized. A phasing mechanism allows the cut to exactly match the printed design. The transmission expression is shown in Scheme 10.1

$$\text{Motor } 1 \rightarrow \text{Shaft I} \rightarrow \frac{\text{Pulley } 2}{\text{Pulley } 3} \rightarrow \text{Shaft II} \rightarrow \text{Reduction box } 4 \rightarrow$$

$$\text{Shaft III} \rightarrow \frac{\text{Gear } 5}{\text{Gear } 6} \rightarrow \text{Shaft IV} \rightarrow \frac{\text{Gear } 7}{\text{Gear } 8} \rightarrow \text{Shaft V} \rightarrow$$

$$\text{Shaft V} \begin{cases} \text{Cutting roll } 13 \\ \\ \dfrac{\text{Gear } 9}{\text{Gear } 10} \rightarrow \text{Shaft VI} \rightarrow \text{Printing roll } 12 \end{cases}$$

Scheme 10.1.   Transmission scheme for a rotary cutter.

The handwheels (14) and (22) are used to regulate the height of the printing and cutting rolls respectively. For example, as the clutch is engaged by pushing down and turning the handwheel (14), the revolving worms (16) and (17) actuate worm wheels (18), (19) so that the screw shafts VIII and IX make the screw blocks (20) and (21), together with the printing roll, be lifted or lowered slightly and smoothly. When proper functioning is achieved, the clutch is disengaged immediately to obviate any misuse of the handwheel.

The capacity of the rotary cutter can be calculated according to equation (10.1).

$$G = 60 \, N_r P \eta_r / K \tag{10.1}$$

where G is the machine capacity kg/h; N_r is the top roll revolving speed in rev/min; P is the number of impressions of the biscuit design on the top roll periphery; η_r is output proportion of perfect biscuits (less than 100%); K is the number of biscuits per kilogram.

Since P and K are fixed and η_r varies only slightly, the best way to increase the machine output is to raise the revolving speed of the top rolls, N_r.

Theoretically, the printing and cutting rolls can revolve at any speed within the limits of the drive. However, the cutting machine is only one part of the automatic biscuit production line, and it should be synchronized with the earlier gauge rolls and the later oven band. Since the dough sheet is so thin that careful handling is needed, too high a speed would induce cracking in the dough sheet. Thus, the rotary cutter speed is dependent on many factors, especially the oven band speed and the top roll diameter. The longer the oven, the greater will be its band speed. The length of the oven is a function of the roll diameter, roll speed, and the baking time. Speeding up of the rotary cutter is therefore limited by these factors.

10.5 MAIN WEBS

Webs play a very important role during the cutting and post-cutting operations for both reciprocating and rotary cutters. Some cutting units have two main webs consisting of a cutting web and a scrap pick-up web (scrap lift web). A single cutting web would eliminate the 'band break' at the scrap lift point by serving both the printing-cutting operation and the transfer of the newly cut dough pieces to the panning unit. Some other machines provide separate webs for cutting and transfer. In these cases, the separating point is just under the nose piece of the scrap pick-up web feed end. 'Having the separate webs helps to release the cut pieces and trimmings from the cutting web prior to transfer onto the panning unit (APV Baker)'. Of course, the cutting web and transfer web are actuated by a common drive so that it is easy to make them run at a matched speed.

By either reciprocating or rotary cutting, the dough is divided into pieces surrounded by scraps. As the dough pieces are transferred for panning and baking, the scraps should be returned to the gauge roll hopper for re-use. Otherwise, the scrap would dry too quickly and not be suited for sheeting. To separate the scraps from the dough pieces, an inclined web, the 'scrap pick-up' or 'scrap lift' web, is placed just after the transfer point between the cutting and transfer web (Figs 8.4, 10.4, and 10.5 for the rotary cutter; Fig. 9.6 for the reciprocating cutter). The pick-up point of the web is just above the height of the dough pieces to allow them to pass through to the panning unit, and it is supported, without exception, by a metal nose piece (knife edge). The web is usually made of woven cotton or food quality synthetic fabric, and it runs upwards at an angle of 20–30°. Occasionally, some plastic-coated materials are used because of their antiwear properties. In any case, sufficient friction between the dough and the web is needed to prevent the scraps, either a continuous 'lattice' or broken lumps, from sliding down the slope. To assist lifting, some machines are equipped with scrap pick-up fingers. However, these can be used only when the biscuit designs are arranged in 'check' rather than in 'brick' patterns.

The scrap pick-up web is usually pivoted above the cutting machine main body to allow the operator to set the optimum transfer position.

At the top of the scrap pick-up web are the scrap/slitter unit which consists of two rotating shafts, one fitted with blades, the other fitted with disks. This unit cuts the dough sheet into rectangular pieces before transferring it to the side scrap return webs (Fig. 9.6).

To prevent travelling off course, all webs are provided with tracking devices including adjustable nose pieces and rolls. Position tensions are applied to the web by weighted rolls and lock down devices. The tensions should be relaxed when not in use, to prevent unnecessary fatigue and to prolong service-life. It is especially important for the cutting web to be away from the top rolls of the rotary cutter when not working.

The webs are usually supported by a stainless steel table which is fixed to the cutting machine sideframes for maximum stability. For food hygiene and successful operation, all webs are equipped with scrapers and scrap trays on their return runs. The trays should be emptied and cleaned and the scrapers should be lowered out of the way when not in use.

Table 10.1   General specfications of some rotary cutters and webs (Courtesy of APV Baker, UK)

Model	Three-roll rotary cutter		Four-roll rotary cutter	
Diameter of printing and cutting rolls	140– 160 mm		140–160 mm	
Diameter of anvil roll	406 mm with 8 mm thick food quality rubber		197 mm dia. with 19 mm thick food quality rubber	
Main drive motor	1.1 kW		1.1 kW	
Dough sheet width (mm)	1000	1200	1000	1200
Net weight (kg)	1970	2032	2000	2100
Dimensions (mm)	1250 × 1774 × 1303	1250 × 1978 × 1303	1350 × 1724 × 1303	1350 × 1928 × 1303
Main web drive motor	Cutting transfer webs: 3 kW max. Scrap pick-up web 1.1 k W max.			

Since all webs should be synchronized with the dough piece cutting unit, they are driven by a variable speed motor through simple mechanisms including driven rolls and an integral reduction gearbox in common.

Table 10.1 shows specifications of the rotary cutter unit and the main webs.

Part V

Cookie-making machines

11

Rotary moulders

Cookies, as the term is used here, means the kind of baked cereal food shaped in one operation without sheeting before shaping. The doughs from which cookies are produced are not suitable for sheeting. According to their formulation and consistency, the doughs can be classified into short and soft doughs. The cookies are classified into short dough cookies and soft dough cookies. The corresponding shaping machines which will be discussed here are rotary moulders for short doughs, and extruders, including depositors, for soft doughs.

11.1 INTRODUCTION

Short dough cookies are related to cake formulations, but they contain much less water. They will cohere under a certain pressure, but readily fracture under tension, for the mixing conditions are chosen to minimize the formation of a gluten network. Owing to their lack of cohesion, such doughs are difficult to handle in the sheeting and cutting process used for the extensible hard doughs for crackers and semi-sweet biscuits. A developed technique has solved this problem. The equipment used in the process is known as a rotary moulding machine, or rotary moulder for short. Fig. 11.1 illustrates a typical machine.

11.2 CONSTRUCTION

Usually, the moulder consists of a stainless steel hopper, a profiled forcing roll, an engraved moulding roll with a scraper, a rubber covered pressure roll, a woven endless extraction web, and a drive motor and transmission system. All are mounted between the plate side frames with control buttons and handwheels.

11.2.1 Dough hopper
The hopper, which is quite small, is attached to the bearing housing of the forcing roll. It is readily removable and it is fitted with a magnetic safety switch.

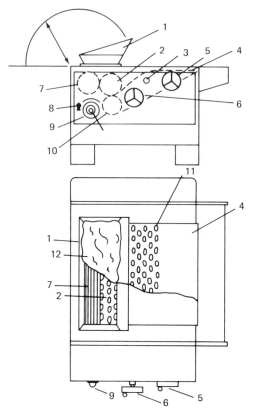

Fig. 11.1 Typical moulding machine (Courtesy of APV Baker, UK). (1) Dough hopper, (2) Moulding roll, (3) Roll speed control, (4) Extraction web, (5) Extraction web tension, (6) Rubber roll extraction pressure, (7) Forcing roll, (8) Electrical isolation switch, (9) Moulding roll scraper adjustment, (10) Rubber roll, (11), (12) Dough.

11.2.2   Forcing roll and moulding roll (Fig. 11.2)

The forcing roll is made of steel and is heavily grooved radially and longitudinally for feeding the moulding roll with short dough materials. Its diameter ranges from 200 to 240 mm.

The moulding roll is of segmented construction. Its centre tube is made of bronze or gunmetal, which are suitable for delicate engraving, or of steel with plastic insert moulds around the periphery to shape the dough pieces. The two steel ends of the moulding roll are fitted onto the centre tube, which carries the shaft on which the appropriate bearings and driving gear wheel are mounted.

To make the negative engravings on the metal centre tube is rather expensive and time-consuming, and it is difficult to repair damage. The plastic inserts make the work much easier and cheaper. In this case, the centre tube is designed with simple holes for the plastic moulds to fit into. To fix them from the inside of the tube seems less easy than with biscuit cutters.

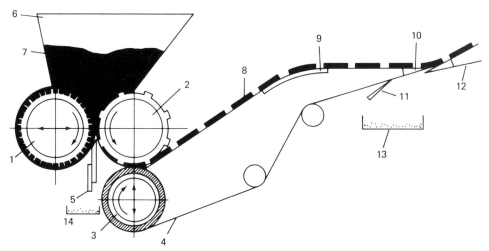

Fig. 11.2   Shaping principle of rotary moulding. (1) Forcing roll, (2) Moulding roll, (3) Rubber roll (extraction roll), (4) Extraction web, (5) Scraper, (6) Hopper, (7) Dough, (8) Dough piece, (9) Adjustable curved support, (10) Nose piece, (11) Scraper, (12) Band to oven, (13), (14) Trays.

The engraved design, its profile, configuration, and depth, are very important factors for satisfactory production. There are many comments on this subject. It is advisable to design the engravings with extensive patterns, since too plain a design tends to have 'suction pad' effect between the dough and the metal. However, too complex a pattern will have a 'keying' effect and will block or clog in the finer parts of the engravings.

The depth and the configuration of the engravings are determined by the requirements of the finished product. But to make the engraving to its correct depth is usually a matter of trial and error. The configurations, either rectangular or round or oval, should be staggered on the moulding roll, for three reasons. Firstly, there is a scraper blade tangential to the moulding roll. If the engravings are in rows the blade would flutter against the roll, as each row of engravings is followed by a ridge which the blade would wear off rapidly to an abnormal condition. Secondly, to maintain uniform stability and strength of the moulding roll requires a staggered configuration. This is especially important for the centre tubes made with holes for the plastic inserts to be fixed into. Thirdly, it allows the forcing roll to feed the moulding roll in a stable condition.

Docker holes are wanted for some cookies, especially for thicker ones, to allow heat penetration more evenly to the centre, and to allow the steam to escape from the underside of the dough pieces.

There are two ways of docker pin making. One is to engrave the docker pins out of the solid brass in the mould impressions. The other is to insert docker pins made from thin brass rod into the plastic moulders.

The length and the shape of the docker pins are also important factors which affect the final product. The tip of the docker pin should be below the surface of the moulding roll, otherwise the scraper blade would catch the docker pin tip, leading to disastrous results. If too far below the surface, blind holes will be produced, causing the dough piece to arch upwards during baking. Therefore, the tapered docker pins should be designed with a rounded end to allow the dough piece to come out smoothly in the extraction process.

The diameter of the moulding roll is a critical factor for the capacity of the machine, which is calculated from equation (11.1).

$$Q = \frac{60 \, NnK}{M} \tag{11.1}$$

where Q is the capacity of the machine (kg/h), 60 is minutes per hour (min/h), N is the number of engravings on the moulding roll (pieces), n is the revolutions of the moulding roll per minute (rev/min), K is the coefficient of the normal operation which is dependent upon the skill of the operator, the dough property, and the conditions of the machine (<1), and M is the number of baked cookies per kilogram (pieces/kg).

Generally speaking, the larger the moulding roll, the more engravings that can be manufactured on it and the more pieces produced each revolution. But the line speed of the moulding roll is dependent upon the oven band speed. So it is usually designed in a standard range (stepless) of 0 to 25 rev/min to match the oven. The diameter of the moulding roll is commonly limited to a range of 200 to 270 mm.

The nip between the forcing and the moulding rolls is adjustable from 5 to 11 mm by altering the position of the forcing roll through two worm gear jacks with a common linkage. The machine is able to retract the forcing roll flanges clear of the moulding roll to assist changeover.

11.2.3   Extraction web and roll

The extraction roll is a compound assembly, a steel cylinder covered by thick food quality rubber. For some machines, the rubber covered roll is driven by the moulding roll through contact with the endless woven canvas web known as the extraction web. For others, it is driven either by an independent motor or by the same motor as the forcing and moulding rolls via a series of mechanical units consisting of a sprag clutch to ensure that the extraction web can be driven only at the same speed as, or slightly faster than, the moulding roll.

From the extraction point to the nose piece, the web should be supported either on a gentle curve, such as an adjustable curved support table or a tensioned slope, to prevent the thick or dried dough piece from cracking. Fig. 11.2 shows the extraction web supported by an adjustable curved table and tensioned by rollers.

The pressure between the moulding roll and the extraction roll via the web is very important for successful extraction of the dough pieces from the engravings. Fig. 11.3 shows a typical extraction roll and the adjustment of the pressure. The rubber covered roll is carried by a static shaft through bearings. The shaft is eccentric and fitted with a worm gear at one end, which is engaged with a worm connected to a handwheel. Turning of the handwheel makes the eccentric shaft revolve so that the height of the extraction roll is adjusted so that the required pressure is obtained. Some machines have an independent device for side trimming.

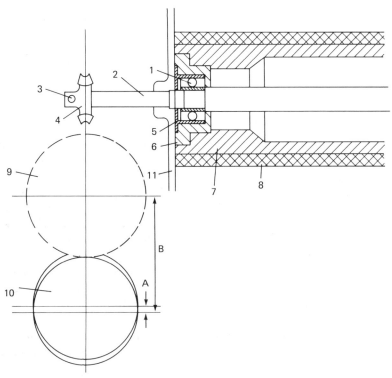

Fig. 11.3   Typical extraction roll (Courtesy of Xin Xiang Food Machinery Factory, China).
(1) Ball bearing, (2) Eccentric spindle, (3) Core pin, (4) Worm gear, (5) Bearing cover, (6)
Bearing seat, (7) Steel roll, (8) Rubber cover, (9) Moulding roll, (10) Extraction roll, (11)
Frame, (A), Adjustable range, (B) Minimum centre distance between moulding roll and
extraction roll.

The lengths of the forcing roll, moulding roll, and extraction roll are designed to
suit the width of the oven band, which is usually 560, 600, 800, 1000, 1200, 1270, or
1500 mm. There is little difference in the diameters of the three rolls. Table 11.1 shows
roll data for four models of rotary moulder.
Modern rotary moulders are equipped with an advanced control system. The
operator controls are mounted on a facia panel on the side of the machine (Fig. 11.1).
Handwheels with colour-coded integral indicators are employed for the adjustment of
the forcing roll gap, scraper position, extraction roll pressure, and web tension. The

Table 11.1   Roll diameters (mm)

Model	1	2	3	4
Moulding roll	200	205	259	260
Forcing roll	200	215	238	260
Extraction roll	200	215	215	267

drive is located at the side of the machine, fully enclosed under lift-off covers which provide total access.

To link with the oven band, the machine speed is infinitely variable within the limits set by the motor speed. The specifications of some modern rotary moulders are shown in Table 11.2.

11.3   THE MOULDING PRINCIPLE

As shown in Fig. 11.2, the rotary profiled roll (1), commonly known as the forcing roll or feeding roll, takes the dough from the hopper and presses it against the reversely rotating roll (2) to fill a series of cookie engravings in the roll periphery. The excess dough is removed by the steel blade of a scraper (5) ('doctor blade') which bears down on the mould, and then is returned to the hopper by adhesion to the forcing roll (1) in the form of a blanket. Between roll (2) and (3) is the endless woven web (4) which is pressed against the moulding roll (2) by adjusting the position of roll (3) in a vertical direction.

The pressing action and the reasonably soft surface of roll (3) allow the web to penetrate into the engravings on the moulding roll (2) and extract the dough. So the dough pieces adhere on the web and are extracted from the engravings on the moulding roll.

The extracted dough pieces adhering on the web are transmitted through the terminal nose piece to another web either for a washing and dusting process or directly into the oven.

11.4   TRANSMISSION SYSTEM

The machine is usually driven by an electric motor through a series of belt and wheel, gear trains, chain and wheels, and a stepless shaft which is essential for an infinite changing of machine speed to suit the oven band speed. Fig. 11.4 illustrates the

Table 11.2   Specifications of some modern rotary moulders (Courtesy of APV Baker, UK)

Model	1		2	
	Forcing roll	Moulding roll	Forcing roll	Moulding roll
Diameter (mm)	238	259	215	205
Speed (m/min)	0 to 25		0 to 12.5	
Length (mm)	800, 1000, 1200, 1250		800, 1000	
Motor (kW)	7.5, 11, or 15		4	
Dimensions (mm)	2163 × 2180 × 1050		2007 × 1530 × 960	
	2363 × 2180 × 1050		1803 × 1503 × 960	
	2563 × 2190 × 1050			
	2613 × 2180 × 1050			
Net weight kg)	3500, 3696, 3840, 3900		1600, 1750	

transmission system, with its expression in Scheme 11.1.

Motor 1 → Shaft I → $\dfrac{\text{Belt wheel 2}}{\text{Belt wheel 3}}$ → Shaft II → Stepless shift 4 → Shaft III

Shaft III → $\dfrac{\text{Gear 5}}{\text{Gear 6}}$ → Shaft IV → $\dfrac{\text{Gear 7}}{\text{Gear 8}}$ → Shaft V

Shaft V → $\dfrac{\text{Gear 9}}{\text{Gear 10}}$ → Shaft VI → $\dfrac{\text{Double chain gear 11}}{\text{Double chain gear 12}}$ → Shaft VII

Shaft VII
$\begin{cases}
\text{Forcing roll 13} \to \text{Dough feeding} \\[4pt]
\dfrac{\text{Gear 16}}{\text{Gear 17}} \to \text{Shaft VIII} \to \text{Molding roll 14} \to \text{Dough shaping} \\[4pt]
\dfrac{\text{Moulding roll 14}}{\text{Extraction roll 15}} \to \begin{array}{l}\text{Dough pieces to be pressed}\\ \text{against the web}\end{array} \\[4pt]
\dfrac{\text{Chain gear 18}}{\text{Chain gear 19}} \to \text{Shaft X} \to \begin{array}{l}\text{Web extracting and conveying}\\ \text{the dough pieces to next}\\ \text{stage}\end{array}
\end{cases}$

Scheme 11.1.   Transmission system for a moulder.

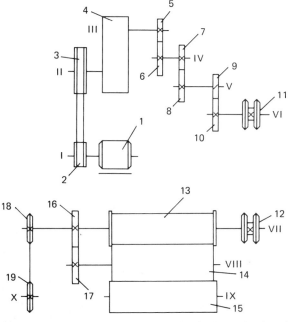

Fig. 11.4   Transmission system (1) Motor, (2), (3), Belt wheels, (4) Stepless shift, (5)–(10) Gear wheels, (11), (12), Double chain gear, (13) Forcing roll, (14) Moulding roll, (15) Extraction roll, (16), (17) Gear wheels, (18), (19) Chain gear. (I)–(X) Shafts.

From motor (1), rotation is transmitted through belt wheels (2) and (3), mechanical shift (4), gear trains, (5–10), double chain gear (11) and (12), and six intermediate shafts I, II, III, IV, V, and VI to shaft VII on which the forcing roll is actuated to force the dough into the engravings on the moulding roll (14).

From shaft VII, rotation is further transmitted in two ways: one is through gear wheels (16) and (17) to shaft VIII which drives the moulding roll (14). By friction, the extraction roll (15) also revolves and allows the web to extract the dough pieces out of the moulders on to roll (13). The other drive is through the chain gear (18) and (19) to shaft X which transmits the torque and motion to the extraction web through other chain and gears.

11.5   OPERATION

The rotary moulding machine simplifies the short dough cookie process. There is no need to form and support a sheet, to gauge, to mould and cut, and to return the cutter scrap dough by a long conveyor. But it is by no means a simple machine in either operation or design. There are many factors related to the final products' quality. Among them are the nature of the dough, the level of dough in the hopper, the gap between the forcing and moulding rolls, the position of the scraper blade, the pressure of extraction, the adjustment of relative speed of forcing, moulding, and extraction roll, and so on.

11.5.1   Nature of dough

For rotary mould cookies, the dough should be relatively dry, even granular, with or without a little elasticity but cohesive enough to allow the dough to stick together if squeezed in the hand so that individual cookies do not tear apart at transfer points.

To maintain steady feeding, the level of dough in the hopper should be substantially constant to reduce pressure differences at the nip and along the whole length of the gap between the forcing and moulding rolls. Too much dough in the hopper will cause not only a large pressure change but also a danger of bridging of dough with a firm consistency. The higher the pressure is, the heavier the dough piece will be, owing to the changes in dough density. In addition, the dough should be carefully examined for metal content to prevent the 'soft' engravings from damage.

11.5.2   Dough piece weight control

Weight changes can be effected by the gap between the forcing roll and moulding roll, as well as the position and the angle of the scraper blade.

The forcing roll, profiled with grooves around its periphery, is adjustable in a working gap range which allows changes of the dough piece weight as the gap is altered. The alteration is usually made by turning a handwheel which drives the eccentric bearing of the forcing roll to obtain a gap range from about 5 to 11 mm.

The position and the setting angle of the scraper blade also effect the dough piece weight changes. The scraper blade should be tangential and as near as possible to the moulding roll, but not so as to dig into the roll. The higher the blade is set, the more the dough is forced into the mould. Conversely, the lower the blade, the less dough will be in the mould. As discussed in section 11.3, when the forcing roll revolves, the dough will be drawn down to the nip, the narrowest place between the centre lines of the forcing and moulding roll, where the maximum dough pressure is exerted. The nearer to this area, the higher the dough pressure will be. Therefore, if the scraper blade is set at a high position, the high pressure dough would be not only pushed by the forcing roll but also squeezed by the blade, which would cause overfilled moulds leading to greater dough piece weights. It is suggested that the blade be set at 3 to 8 mm under the axis of the moulding roll, while the standard issued by the Ministry of Commerce of China stipulates that the blade must be set at 2 to 5 mm under the moulding roll axis. In Fig. 11.1, the moulding roll scraper adjustment is carried out by the handwheel (4) with a colour-coded integral indicator.

11.5.3 Dough piece shape control

The shape of the dough piece can be affected by the pressure brought by the rubber covered roll at the extraction web/moulding roll interface and the speed of the web in relation to the moulding roll speed. As the dough piece is brought into contact with the web, it sticks onto the web first because of the pressure and elasticity of the rubber covered extraction roll, the roughness of the web and the adhesiveness of the dough. As soon as the dough piece leaves this interface, it is extracted by the web because of the stronger adhesiveness of the dough to the web compared with the engraving roll, which is usually coated with Teflon to prevent sticking. However, the speed of the extraction web warrants further consideration. It can be driven only at the same speed as, or slightly faster than, the moulding roll. Otherwise, it would tear the dough out of the engraving too quickly or squeeze the dough into a 'fin' (or tail) at the back or into other cookie shapes at too low a speed. It must therefore be in harmony with the moulding roll speed. To achieve this, a sprag clutch is fitted within the drive mechanism for the rubber extraction roll to ensure synchronization. The extraction web tension, rubber roll extraction pressure, and roll speed, can all be easily adjusted by the corresponding handwheel with indicator (Fig. 11.1).

12

Extruder and depositor

12.1 INTRODUCTION

Danish batter cookies, Viennese whirls, and Spritz cookies are rich in fat, brandy snaps are rich in sugar, and Jaffa cakes, sponge boats, and lady's fingers are sponge textured. The doughs of the above-mentioned foods are either relatively viscous, fruited, or have small particle ingredients included or are soft enough to be just pourable. These types of dough are conveniently referred to as soft doughs. The formation of this kind of dough cannot be carried out by sheeting, printing, rotary cutting, or even rotary moulding because their rich fat, or fatless, aerated, or coarse texture and rheological properties with the associated high viscosity and flowability make it difficult to shape and to draw from moulders.

Extruding machines are expressly designed to deal with the formation difficulties of soft dough. 'Depositing' is a form of extrusion, so these two methods of dough piece forming are not distinct from one another [11]. Hence, the extruder and depositor will be discussed together in this chapter. Fig. 12.1 shows a front view of a typical extruding machine (depositor).

12.2 CONSTRUCTION

A typical extruding machine consists of a dough feed assembly, die assembly, power and transmission, frame, and control system.

12.2.1 Dough feed assembly

There are two different types of dough feed assembly: a grooved roll feeder, and a plunger pump feeder. In either case, a stainless steel hopper is always needed to allow the dough to be drawn either by two grooved rolls or by a row of plunger pumps and further forced into the die underneath. The dies are arranged in a row across the width of the band.

12.2.1.1 Grooved roll feeder

In this feeder, the feed rolls, fitted with adjustable scrapers, are made of high duty iron with longitudinal grooves, and are mounted in parallel under the dough hopper. The gap between them can be adjusted by a handwheel via a worm and wheel mechanism.

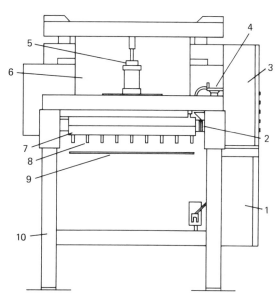

Fig. 12.1   Extrusion (depositor) machine (Courtesy of APV Baker, UK) (1) Gear box, (2) Feed assembly, (3) Control panel, (4) Feed adjustment handwheel, (5) Shaping head up–down mechanism, (6) Dough hopper, (7) Die plate, (8) Die nozzle, (9) Receiving web, (10) Frame.

Usually, the rolls are 200 to 300 mm in diameter and 560 to 1250 mm in length to match the oven band width. For some extruders, the feeding rolls are driven by only one motor, while in others each feed roll is driven independently by a variable speed motor and worm gear reduction box.

Comparatively, the grooved roll is easier to manufacture than the plunger pump feed assembly, and the machine is also simplified by using the roll feeding system. However, rolls with different grooves should be used for doughs with different consistencies, such as the doughs for Viennese whirls and for sponge boats, which are different from each other. Fig. 12.2(a) and (b) shows two types of feeding roll assembly for a viscous even coarse textured dough (a) and a pourable dough (b) respectively.

The construction, the method of installation, and the rotary direction of the grooved rolls differ from one another when dealing with doughs of different consistency. During the feeding of unpourable doughs, the rolls are rotating face to face, and the dough is forced downward into the die through the nip between the two rolls. With pourable doughs, the rolls are rotating reversely and the dough is forced to go along the side of the chamber wall and then into the die underneath. In either procedure the rolls always apply force to the dough at the hopper bottom. Because of this, the rolls are also referred to as forcing rolls, as in the rotary moulding machine. Between the dough hopper and the die assembly is the dough chamber, in the lower part of which the pressure is balanced as the pressure is increased with the entering dough from the hopper, but decreased with the expelling of dough through the die. The chamber is often known as a pressure balance chamber.

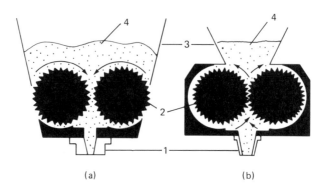

Fig. 12.2   Grooved roll feed systems (a) for viscous dough (b) for pourable dough (1) Filler
block, (2) Grooved feed roll, (3) Dough hopper, (4) Dough.

12.2.1.2   Plunger pump feeder

In this feeder, a row of small plunger pumps is employed, each of which consists of one cylinder and one piston, and corresponds to one die.

Compared with the grooved roller, the pumps are not so easy to manufacture and the machine is more complicated in structure. The advantages of this feeder are in the easy controlling and measuring of the size and weight of the products, as well as a wide usability for doughs ranging from very high consistency to very thin (easy pourable) materials, so that it can be used for making an enormous variety of products. An extrusion machine with a pump feed assembly is often referred as a depositor, for its best application is the depositing of sponged and aerated pourable doughs.

As shown in Fig. 12.3, the feeding procedure has two stages, suction and extrusion. In Fig. 12.3(a), the plunger (piston) (7) is moving leftward in its suction stroke within the cylinder (6) so that the empty space left by the plunger is in vacuum, causing the dough to be sucked in (drawn down) through the passage formed by the valve (2) between the dough hopper and the cylinder. The valve is mounted on a cylindrical shaft partly milled off to form a series of slots of which each corresponds to a feed cylinder. Driven by a rack-pinion mechanism, the shaft positively and reversely revolves by 90° intermittently, so that the valves swing in step, as the shaft does, to open or shut the passage between the feed cylinder and the extruding die.

In Fig. 12.3(b), the valve has already rotated 90° anticlockwise so that the passage between the dough hopper and the feed cylinder has been shut and another passage between the cylinder and die through the filler block (1) has been opened, as the plunger has already finished its suction stroke and begun its return stroke by moving rightward. Along with the movement of the plunger, the pressure in the right part of the cylinder is increased. Since the passage to the die has been opened by the valve, the dough pushed by the plunger has to go downward through the filler block and is extruded out of the die.

As soon as the preset amount of dough is extruded from the die, the plunger stops its return stroke and begins to move leftward in its suction stroke while the valve turns

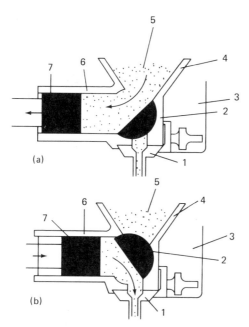

Fig. 12.3   Plunge pump feed system (a) suction, (b) extrusion (Courtesy of APV Baker, UK)
(1) Filler block, (2) Valve, (3) Shaping head, (4) Dough hopper, (5) Dough, (6) Cylinder, (7)
Plunger.

by 90° clockwise so that the passage to the die is shut and another passage from the
hopper to the cylinder is opened, leading to the next feeding cycle.

By using the plunger pump feeding assembly, the machine has performed its most
useful application as a depositor for either sponged or aerated pourable doughs. It
provides easy control of unit product weight by adjusting the plunger's travel.
Technical specifications of a typical depositor are given in Table 12.1.

12.2.2   Die assembly

This assembly consists mainly of a set of individual dies fixed on a die plate (9) which
is connected with the filler block (3) by means of the clamps (10). There are a series of
holes on the die plate, each hole corresponds to a die. The filler block is mounted at
the base of the dough chamber to introduce the dough fed by the forcing rolls or
cylinder pumps to the die through the die plate for extruding.

The die can have a wide range of sizes and shapes (Fig. 12.4) to form various dough
pieces. The die is not universal, since the texture and consistency of the doughs for
extruding differ from each other; some doughs are suitable for a wire-cut, some for a
press, while some for depositing. Therefore the die should be selected carefully, before
operation, to suit the properties of the dough for satisfactory production. The
production of different foods and the selection of their dies will be discussed in section
12.3.

Table 12.1   Depositor specification (courtesy of APV Baker, UK)

Total deposit weight:	113 g to 1130 g per stroke of specific gravity 1.0 Maximum weight is *pro rata* with a batter of lower specific gravity
Speed:	10–30 stroke/minute
Pitch (maximum):	190 mm
Draw (maximum):	152 mm
Maximum weight on table:	20 kg
Pan width:	400 mm to 500 mm
Hopper capacity:	45.4 litres
Drive motor:	0.75 kW (1 hp)
Machine weight:	Net 620 kg
Maximum pan depth:	50 mm, using long dies 100 mm, using pinned short dies

For unpourable doughs, the die can be designed in a variety of shapes, and even on a revolving mode, by being fitted to a gear ring driven by a gear rack. As a result of ingenious design in the application of various die shapes and arrangements, as well as for compounding the movement of the die and the receiving web (rectilinear and rotary, spacing and continuous), many different products can be made by a single machine.

12.2.3   Drive and control
The drive components of the machine are commonly fixed on plate stretchers between the sideframes, clear of the floor. The machine frames and covers are usually made of mild steel with a painted finish. All moving parts are mounted behind steel covers with access panels.

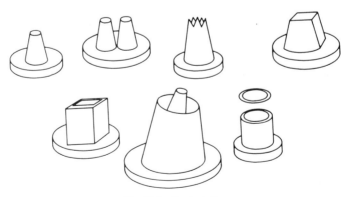

Fig. 12.4   Typical dies.

For food hygiene, all components in contact with the food product are made of either stainless steel or are nickel-plated. Lightweight scrap trays for feed roll scrapers are provided and are easy to remove for cleaning to prevent dough spillage onto the oven band.

An auto/manual selector switch is provided. When switched to the 'auto' position, the feed rolls will follow the wire cut mechanism in cascade. When switched to the 'manual' position, the cascade system is not implemented, and all speed settings are independent of each other (APV Baker, UK).

12.3 SHAPING PRINCIPLE

12.3.1 Shaping of rout press products
The doughs for this type of product are thinner or batter-like. In this application, the dough pieces are produced by extruding the dough through a die which has a series of slit openings approximate to the cross-section of the cookies. The extrusion is a continuous ribbon, and the dough pieces are cut to length by a separate motor driven guillotine cutter before or after baking. Fig. 12.5 shows the shaping principle.

For rout press products, the forcing rolls, as shown in Fig. 12.5, are rotating face to face continuously; the dough is forced downward from the nip to the die and then extruded onto the moving web. The appearance of the upper surface of the extruded ribbon can be altered by changing the nozzle shape, which can be round, rectangular, or with a sawtoothed end to make ridges or lines on the top of the dough pieces. The flow rate of the dough is mainly dependent upon the forcing roll speed and the cross-section of the nozzle. Hence, changing the feeding speed and the size of die can give a wide variation in machine capacity.

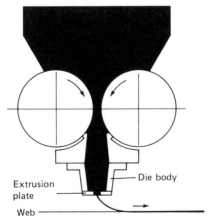

Fig. 12.5   Rout press products (Courtesy of APV Baker, UK).

12.3.2   Shaping of filled bar products

As shown in Fig. 12.6, in this process the upper hopper and the lower chamber are divided into two sections by a division plate. One section contains the filling materials, such as a fig or date paste, which is fed by the action of one of two grooved rotating rolls through the tubes in the die assembly to form the centre part of the products. The other section of the hopper contains the casing dough, which is fed by another feed roll. The ratio of the filling to casing is controlled by varying the diameter of the inside tube and the feeding speed of the independently driven rolls, and it can also be adjusted by the slide plate at the die opening. Like the rout press product, the filled product is extruded in a continuous ribbon, which will be cut to the preset length before or after baking by a guillotine or a rotary cutter driven by a separate motor and gears.

In the production of filled bar products, the machine is also in a continuous mode, as in Fig. 12.5. The feed rolls rotate face to face continuously, and the web moves in a single direction to the oven uninterruptedly. Between the die and the web, there is no relative movement in the vertical plane. But the conditions are different when the machine is used for the production of sponge batter drops.

12.3.3   Shaping of sponge batter drops

The production of sponge batter drops can be carried out successfully by means of a depositor with a plunger pump feed assembly for intermittent feeding and easy weight-controlling characteristics. The shaping process and relative motions between the die and web (tray) will be described together with the transmission system in section 12.4. Here, the shaping process is performed by an extruder with a pair of grooved feed rolls (as in Fig. 12.5 and 12.6). Fig. 12.7 illustrates the shaping principle for sponge batter drops.

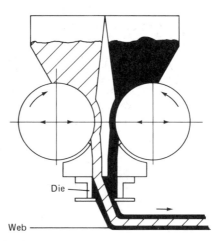

Fig. 12.6   Filled bar products (Courtesy of APV Baker, UK).

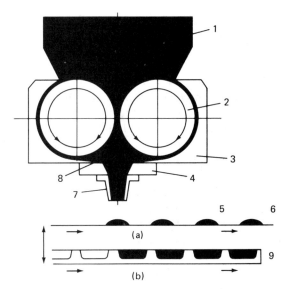

Fig. 12.7   Sponge batter drops (1) Dough hopper, (2) Feed roll, (3) Dough chamber, (4) Filler block and die plate, (5) Sponge batter drop, (6) Oven band, (7) Die nozzle, (8) Pressure/ balance chamber, (9) Mould plate (tray).

Since the sponged batter is aerated and pourable, the feed rolls are installed in a different way to allow the dough to go sideways along the chamber walls, and the grooves on the roll surface have different shapes (Fig. 12.2(b)).

The diameter of the conical nozzle (die) used for depositing is less than that of other kinds of die to allow more precise control of the weight of the dough piece and to produce smaller button-like drops on the tray or oven band.

In this process, the deeply-grooved feeding rolls rotate back to back to feed the batter through the side channels to the chamber underneath. When the batter is fed and extruded from the nozzle, the oven band or tray moves up close to the nozzle to receive the deposited drops. As soon as the preset amount of batter is deposited, the feed rolls should reverse immediately so that an instant vacuum occurs in the die to make the dough go back up but not drop again while the band or tray moves down away from the nozzle and moves one step (equal to the space of about one pitch) towards the oven. To maintain a relatively uniform weight and shape of the products, the variation of the rotary feeding rolls should be synchronized with the movement mode of the receiving band.

If the tray is smooth, the deposited drops will be naturally formed into round thinner pieces with a spherical top by surface tension.

To make thicker pieces, the receiving tray should be designed as a mould plate with round or square or rectangular or any other desired configurations.

As suggested by D. Manley in *Technology of biscuits, crackers and cookies*, 'Two depositors working in unison allow jam to be dropped on top of the batter before baking. The jam reduces the lift in the centre of the sponge in the oven, and this is beneficial when it comes to packaging as the sticky jam zone is recessed so does not come into contact with the base of another piece'.

12.3.4   Shaping of wire cut products

For a rather sticky dough and dough containing coarse particles, such as nuts or oatflakes, a wire cut assembly is needed (Fig. 12.8). In this process the division plate (as used for filled bar products) is removed, the gap between the two feeding rolls is adjusted to allow the sticky or coarse textured dough to go down through the nip, and a wire cut die finger assembly and steel fabric frame (support) (5) with fingers (11) is fitted with die apertures. The steel wire (piano wire or blade) (4) is stretched across the frame and is held taut by a turnscrew.

The wire cut dough piece is produced in two stages. Firstly, the dough is extruded, through a group of die openings under pressure from the feed rolls rotating face to face, into the bar. Secondly, as the bar gains the predetermined length, it is cut off by a piano wire or a blade with a sharp smooth edge or a serrated edge. The wire acts in a reciprocating mode; it is close to the die in its cutting stroke but moves apart from the die in its return stroke to avoid destructive interference to the continuously descending dough extruded by the feed rolls. The cutting stroke direction of the wire can be in the same direction as the oven band, that is in a forward motion. But it is usually in the opposite direction as shown in Fig. 12.8.

The peripheral shape of wire cut products is dependent upon the die (nozzle), which is very simple (round, square, rectangular, or any other shape without sharp angles) for the stickiness or coarse texture of the dough, as well as for the simple wire cut process.

The stroke length of the wire is mainly dependent upon the dimension of the die in the moving direction of the band. The amount of wire drop depends on the thickness of the products, to ensure enough clearance on the return stroke for the newly

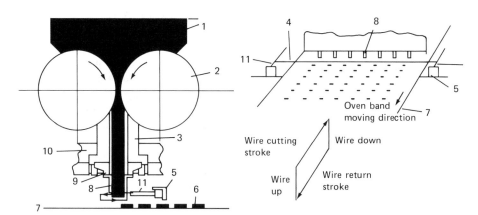

Fig. 12.8   Wire cut products (1) Dough hopper, (2) Feed roll, (3) Filler block, (4) Wire, (5) Wire support, (6) Dough piece, (7) Pan or oven board, (8) Die, (9) Die plate, (10) Die plate and filler block clamps, (11) Finger.

extruded dough. The drop amount and the stroke length of the cutting wire can be adjusted by the corresponding handwheel with its integral indicator.

The distance between the die and the band is important for the dough piece placement as it drops onto the oven band. If the die is too far from the band, it will cause the cut dough piece to be out of shape as it falls on the band. Conversely, if it is too close to the band it will make the wire cutting difficult. A suitable gap is achieved by a handwheel raising or lowering the band directly beneath the die for optimum dough piece placement.

The wire cutting assembly is driven by a variable speed motor via a spur gear reduction box, while each of the two feed rolls is independently driven by a variable speed motor and worm gear reduction box. The machine is therefore usually equipped with three drives which are linked electrically to synchronize their speed. For filled bar and rout press products, the wire cut drive is switched off. The variable speed motors and mechanical reducers make it possible to follow any speed alteration of the oven band or tray. Table 12.2 lists specifications of the extruder.

12.4   TRANSMISSION SYSTEM

Fig. 12.9 illustrates the transmission system of an extrusion machine with a plunger pump feeder. Its expression is shown in Scheme 12.1. Unlike the one equipped with the

Table 12.2 Specifications of some modern extruders with grooved roll feeder (Courtesy of APV Baker, UK)

Model	1			2	
Roll diameter (mm)		297			216
Band width (mm)	1000	1200	1250	800	1000
Machine net weight (kg)	2995	3350	3455	2544	2845
Roll gap (mm)		7.5		7.5	
Wire drop			Max. 19 mm		
Wire stroke			Min. 20 mm Max. 140 mm		
Roll drive motor			5.5 kW		
Wire cut drive motor			Max. 4.0 kW		
Wire oscillating drive motor			Max. 1.1 kW		
Dimensions (mm)	1573 × 1600 × 2015	1773 × 1660 × 2015	1823 × 1600 × 2015	1362 × 1600 × 2000	1562 × 1600 × 2000

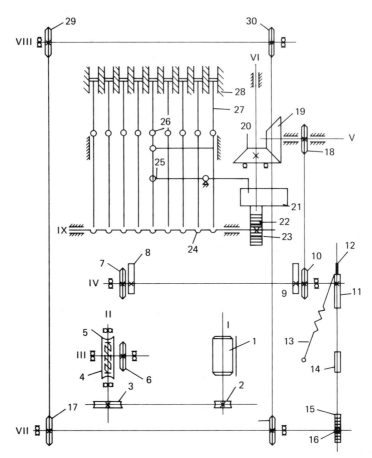

Fig. 12.9   Transmission system [15] (1) Motor, (2), (3) Pulleys, (4) Worm, (5) Worm gear, (6), (7) Chain wheels, (8), (9) Cams, (10) Chain wheel, (11) Cam, (12) Pawl, (13) Return spring, (14) Adjusting nut, (15) Ratchet wheel, (16), (17), (18) Chain wheels, (19), (20) Bevel gears, (21) Cylindrical cam, (22) Rack, (23) Pinion, (24) Valve, (25) Swing lever, (26) Universal joint, (27) Plunger, (28) Cylinder, (29), (30) Chain wheels (I)–(IX) Shafts.

grooved feeding rolls, this machine is driven by only one variable speed motor. From motor (1), the rotation is transmitted through reducing pulley (2) and (3) to shaft II from which the worm gearing (4) and (5) transmits the rotation to shaft III. Actuated by chain gear assembly (6) and (7), shaft IV transmits the power via three routes.

(A)   Cams (8) and (9) on shaft IV drive the shaping head up and down. The travel is about 30 mm. The shaping head is a combination of the dough hopper, pump feed assembly, and die assembly. The shaping head goes downward to extrude the dough onto the receiving band as the die is closer to it. The head returns back (goes up), leaving a clearance to allow the band to move forward one step. For easier understanding, Figs 12.1 and 12.3 will be helpful.

Motor 1 → Shaft I → $\dfrac{\text{Pulley 2}}{\text{Pulley 3}}$ → Shaft II → $\dfrac{\text{Worm 4}}{\text{Worm gear 5}}$ → Shaft III

Shaft III → $\dfrac{\text{Chain wheel 6}}{\text{Chain wheel 7}}$ → Shaft IV

Shaft IV $\begin{cases} \dfrac{\text{Cam 8}}{\text{Cam 9}} \to \text{Shaping head (moving up and down)} \\[2ex] \text{Cam 11} \to \dfrac{\text{Pawl 12}}{\text{Ratchet wheel 15}} \to \text{Shaft VII} \\[2ex] \dfrac{\text{Chain wheel 10}}{\text{Chain wheel 18}} \to \text{Shaft V} \to \dfrac{\text{Bevel gear 19}}{\text{Bevel gear 20}} \to \text{Shaft VI} \end{cases}$

Shaft VII → $\dfrac{\text{Chain wheel 16,17}}{\text{Chain wheel 29,30}}$ → Oven band(tray) (moving step-by-step)

Shaft VI → Cylindrical cam 21

Cam 21 → $\begin{cases} \text{Upper cam-follower mechanism} \to \text{Swing lever 25} \to \text{Plunger 27} \\ \hphantom{\text{Upper cam-follower mechanism}} \text{(sucking and extruding)} \\[2ex] \text{Lower cam-follower mechanism} \to \dfrac{\text{Rack 22}}{\text{pinion 23}} \to \text{Shaft IX} \end{cases}$

Shaft IX → Valve 24 (positively and reversely revolving by 90°
to form inlet and outlet passage for dough)

Scheme 12.1.   Transmission system of an extrusion machine.

(B)   Cam (11), which is T-grooved, through pawl (12) and ratchet wheel (15), drives shaft VII to revolve step-by-step. From shaft VII, the motion is transmitted through chain gears (16), (17), (29), and (30) to the oven band or tray (pan) which is also moving step-by-step to match the shaping head mode.

(C)   From shaft IV, the rotation is transmitted through the chain gearing (10) and (18) to shaft V on which the bevel gear (19) rotates at a speed the same as that of shaft IV. Through the spline shaft VI and bevel gear (20) the cam (21) is actuated. This cam looks like a short cylinder, but there are two grooves on its inside wall which form two inner cylindrical cams. By means of the swing lever (25), the upper cam-follower mechanism of cam (21) makes the plungers (27) travel reciprocally so that dough suction and extrusion are achieved. The lower cam-follower mechanism of cam (21) actuates the valves (24) which are on shaft IX. They revolve by 90° reciprocally by means of rack (22) and pinion (23).

Before the plungers (27) begin their suction stroke, valves (24) on shaft IX have already cleared the dough inlet passage (from the hopper to the cylinder). Likewise, before the plungers begin their return stroke, the valves have already reversed 90° to shut the inlet and to clear the outlet passage for dough expulsion. By that time, the plungers have expelled the dough into the dies, from which the dough is extruded and falls onto the receiving band (or tray). The shaping head, driven by the cams (8) and (9), has already arrived at its lowest position to meet the requirement for a short distance between the die and the band.

After each dough piece drops, the shaping head always moves back up to its top position by means of cams (8) and (9) on shaft IV. Along with the shaping head the dies are away from the band, which moves towards the oven one step to leave a clearance under the die for the next cycle.

Scheme 12.1 is the transmission expression of the machine.

12.5   OPERATION

12.5.1   Weight control

It is much more difficult to control the weight of the products in the extrusion and depositing process than in the sheeting and cutting process. Of course, the weight precision is mainly dependent on the forming machine, but the uniformity of the dough texture is also a key factor to evaluate. Unlike sheeted dough, the dough for extrusion is usually either very viscous, fruited, or of coarse texture, and it is shaped only after mixing. To improve the uniformity of the dough, a thorough mixing and a resting period (half an hour is common) are needed to supply the extrusion machine with a uniform dough not only in texture but in rheological properties. In a certain period, the amount of dough extruded depends mainly on the pressure from the back of the die (the pressure/balance chamber). Usually, the pressure is controlled by the speed of the feeding rolls. But the height of the dough in the hopper also has the potential to affect the pressure. Therefore it is necessary to keep the dough at a relatively stable level in the hopper. The practice of trial and error shows that it is easier to control when the dough is at a lower level.

However, even if the abovementioned measures are employed, non-uniform extrusion weight sometimes happens. In most cases, the side piece is slightly lighter than the centre one. This can be adjusted by a separate mechanism, but it would make the machine more complicated and the design inconvenient. Comparatively, the plunger pump feed assembly demonstrates its advantages in weight control. In Fig. 12.9, weight control can be carried out by adjusting the plunger travel between 1.5 and 35 mm. As the fulcrum of lever (25) moves leftward the radius of the lever vibration becomes larger, so that the plunger travel lengthens, which results in an increased amount of dough both in suction and expulsion, leading to heavier products. Conversely, lighter products will be produced.

12.5.2   Changing the die plate and filler block

For cleaning purposes or for different shaped products, the die plate and filler block should be disassembled and changed. As shown in Fig. 12.8, the dies (8) are fixed on the die plate (9) which is retained by the die plate and filler block clamps (10). The filler block (3) fits into the roll nip to supply a constant volume of dough to each die cup. To dismount this assembly, the die plate can be lowered by an elevator mechanism as the clamps (10) are released by using a dual purpose handwheel in which the two functions (elevator and clamps) are combined. The two operations are achieved by using the shift knob located in the centre of the handwheel.

As the die plate is lowered to band level, it can be withdrawn from the control side of the machine. By first raising the elevator complete with carrying plate, the filler block can then be lowered and removed on the carrying plate through the same aperture as the die plate after disengaging the latches.

Part VI

Cake-making machines

13

Encrusting machines

13.1 INTRODUCTION

Japanese 'manju', western cakes such as 'pies' and Chinese 'mooncake' are stuffed foods. The crust can be yeast-raised or chemically-raised breadlike doughs, hard doughs, short doughs, or other viscoelastic or plastic food materials. Inside the crust is a filling which can be jam, diced fruit or vegetable, or meat, minced or even cooked in large pieces. Most of these types of food have a traditional high moisture and soft tactile sensation.

The early encrusting machine was made like a multi-die press with a round rotatable working table on which about eight dies were fixed in a circle. Above the dies, the corresponding punches were made to reciprocate vertically by a step-by-step mechanism. The formation of an encrusted food was performed in about eight steps. Firstly, a deep hollow was drawn in the dough piece in three steps; secondly, the hollow was filled with stuffing; thirdly, the opening was sealed in three steps; fourthly, the surface was printed with a pattern. When the working table revolved another step, the product arrived at the outlet, where a web led the formed piece to the oven. The baked food made in this way seemed unpopular because of its lack of a soft sensation. The general opinion is that the direct hollow-punching and simple closing-up process made the dough texture too dense, and it was accompanied with a greater stress.

The Rheon encrusting machine solved these problems perfectly. 'Rheon' is derived from the word 'rheology'. Food rheology is the study of the flow and deformation of food and its materials. The Rheon encrusting machine applied the science of rheology engineering, in skilfully manipulating the viscosity and elasticity of dough, to derive a delicate and natural flavour and taste and a satisfactory 'feel'.

The machine has been designed and manufactured to meet the requirements of food processors who cater for convenience food markets, commissaries, snack bar chains, restaurants, and baker's shops. Its revolutionary operating principle is equally applicable to produce uncooked, oven-baked, steamed, fried, and frozen products of every type in the bakery, confectionery, and snack field. Since the first unit came out in 1963, this automatic machine has been developed greatly and successfully into the unique Rheon Model N208. Table 13.1 gives its specification.

Table 13.1   Specifications of Rheon 208 encrusting machine (Courtesy of Rheon Automatic Machinery Co. Ltd Japan)

Model			N208- Type 20	N208- Type 30	N208- Type 40
			Type 20	Type 30	Type 40
Encrusting disk	Moulding speed		20 pcs./min	30 pcs./min	40 pcs./min
	Product weight		10–300 g/pc.	20–120 g/pc.	15–90 g/pc.
Flourless encrusting unit	Moulding speed	Normal speed type	20 pcs./min	30 pcs./min	40 pcs/min
		Double speed type	40 pcs./min	60 pcs./min	80 pcs./min.
	Product weight	Normal speed type	20–80 g/pc	20–80 g/pc.	15–80 g/pc.
		Double speed type	10–40 g/pc.	10–40 g/pc.	7–40 g/pc.
	Weight			550 kg	
	Motor			2.6 kW, 3 phase (SS Type 2.2 kW)	
Dimensions (mm)	208SS type 1845 × 1065 × 1457		208SD type 2055 × 1065 × 1457	208DD type 2416 × 1065 × 1457	

13.2   CONSTRUCTION

13.2.1   General structure

Since it performs dividing, rounding, moulding, and filling automatically and efficiently, the machine is relatively complicated but very compact, occupying an area of less than 2 m². Its electrical consumption is only 2.6 or 2.2 kW. As shown in Fig. 13.1, the machine consists of a main body mounted on castors and levelling jacks which allow the machine to be moved and installed easily. Two encrusting disks and an up and down stability disk, work conveyor, and flour dusting unit are mounted on the main body. The casing and filling feed assemblies are mounted on a turret which can be swung sideways for adjustment, cleaning, and maintenance. A clamp locking lever holds the turret in position. The feed assemblies have two hoppers: the side hopper for the casing material, and the centre hopper for the filling material. The two different materials are impelled by two sets of counter-rotating screws, and are formed into a cylindrical shape at the compound nozzle section.

The machine is equipped with a 2.6 kW (2.2 kW for SS-Type), 3-phase motor and two gearboxes which give the choice of ten different ratios of filling to casing. These are controlled externally by means of a clutch lever for each box.

There are three versions of the Rheon 208 encrusting machine which differ with respect to their work conveyors. The SS type is equipped with one option base and a two-step speed change mechanism; the SD type has a one option base and a stepless speed change mechanism; the DD type has a two option base, a stepless speed change mechanism, and two belt conveyors arranged in tandem.

The option base in the conveyor section is for driving the movable secondary shaping options and for synchronizing with the conveyor speed. The movable shaping options have various types of shaping and working tools corresponding to the need to eliminate labour intensive handwork required after encrusting with the special

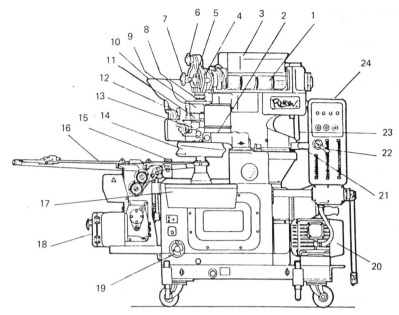

Fig. 13.1   Rheon encrusting machine (side view) (Courtesy of Rheon Automatic Machinery Co. Ltd, Japan) (1) Centre cylinder, (2) Dusting flour unit, (3) Centre hopper, (4) Filling controller, (5) Worm, (6) Worm gear, (7) Worm, (8) Filling guide, (9) Chuck, (10) Side hopper, (11) Turning nozzle, (12) Side clamp, (13) Compound nozzle, (14) Encrusting body, (15) Option base, (16) Work conveyor, (17) Flour receptacle, (18) Conveyor speed control, (19) Handwheel for adjusting height of stability disk, (20) Motor, (21) Turret locking lever, (22) Main switch, (23) Centre clutch, (24) Switch box. I, II, III Clamp shafts.

shaping of many different products. By combining the operation of one or two operational tools and two conveyor belts with individual speed adjustment, more complicated secondary shaping is possible.

To prevent damage to the machine, all main drives are fitted with shear pins which automatically break and stop turning when an abnormal situation occurs.

The lubrication systems are designed to oil the gear boxes via the oil inlets, and they provide oil gauge inspection.

The electrical control box is fixed on a free-standing conduit clear of the main body.

13.2.2   Feed system
Since the product is composed of two different materials, the machine is equipped with two sets of feed assemblies to transfer the casing and filling materials into the inner and outer nozzles separately. Fig. 13.2 is a vertical view to show the working position of the two feed assemblies.

13.2.2.1   Casing material feeder
This is also called 'side feeder' as its working position is on the left side of the machine. The unit comprises two conveying screws assembled in parallel in the cylinder which

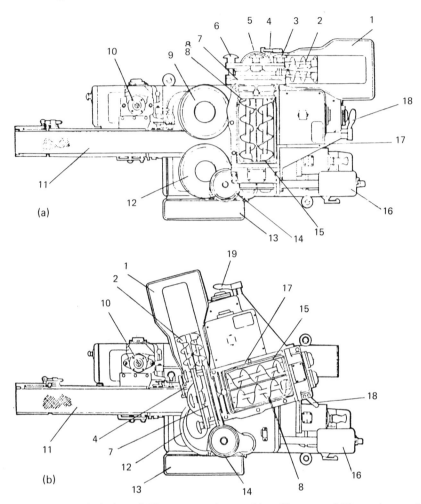

Fig. 13.2 Vertical view of Rheon encrusting machine (Courtesy of Rheon Automatic Machinery Co Ltd, Japan) (1) Side hopper, (2) Side screw, (4) Side clamp, (7) Filling controller, (8) Centre screw, (9) Left encrusting disk, (10) Option base, (11) Work conveyor, (12) Right encrusting disk, (13) Flow receptacle, (14) Flow dusting unit, (15) Centre hopper, (16) Switch box, (17) Centre cylinder, (18) Centre clutch.

is called the 'side cylinder' to distinguish it from the filling feeder cylinder. The side cylinder is placed in its frame where the right end supports the side cylinder opening against the compound nozzle to allow the casing material to pass through into the outer turning nozzle. At the left end of the frame is a small gear box to transmit rotation to the two side screws (18) (Fig. 13.6) through end-on couplings. The screws should be set in the correct way according to the different shape of the couplings, otherwise the threads will not roll together (Fig. 13.6).

There are three available types of screw. The standard one is the forced-type which is applicable to almost all sorts of casing material. A second type is the stainless steel screw fitted for minced meat, fish paste, and similar materials. The last type is a blind shaft which is set into the side cylinder to replace one of the two screws to reduce the casing output by half when producing small products.

The stability roller is positioned above the side screws to apply pressure to the casing material so that it engages with the side screws to ensure stable feeding. This roller is especially helpful when feeding yeast-raised doughs.

The side hopper is fixed on the side cylinder frame to introduce the casing material into the machine. It can be easily locked by turning the clamp lever downward (Fig. 13.5).

13.2.2.2   Filling material feeder
Since the filling material is always in the centre of the product and is fed to this position , the filling material feeder is called the 'centre feeder'. It comprises the centre hopper, centre cylinder, centre screw, filling controller, etc.

As with the side feeder, two centre screws are set in the centre cylinder, and they receive power from couplings on the centre cylinder frame. Since the filling materials used are often visco-plastic, the stability roller is often omitted and the centre controller must be used to deliver the filling material into the inner turning nozzle.

The filling controller is a vane or paddle pump with two sets of wings which prevent the filling from shearing or being subject to pressure, so that even the most delicate textured fruit fillings can be introduced without spoiling their consistency (Fig. 13.3

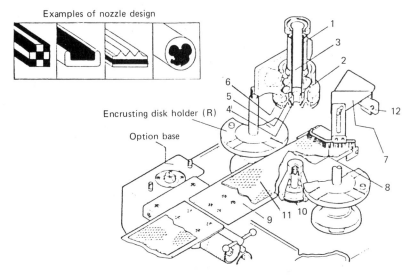

Fig. 13.3   Extrusion nozzle moulding (Courtesy of Rheon Automatic Machinery Co. Ltd, Japan) (1) Outer nozzle, (2) Extruder compound nozzle, (3) Inner nozzle, (4) Nozzle tip, (5) Fixed ring, (6) Retaining ring, (7) Flour distributor, (8) Brush, (9) Back plate, long, (10) Junk cap, (11) Conveyor belt, long, (12) Distributor bracket.

and 13.6). The wings revolve along the special curved wall of the filling controller case, causing the filling material to change its direction of movement by 90° and to move downward into the turning nozzle (Fig. 13.5) or into the inner nozzle (Figs 13.3 and 13.4) through the filling guide at a preset rate. The working principle of this kind of filling controller has been discussed in Chapter 4.

The filling controller case is covered by a transparent plate to allow clear observation of the feeding process. There are two air cocks on the case cover to expel air pockets, otherwise they can cause irregular feeding of the filling material. To avoid any leakage, the cover is fastened tightly by means of three clamp shafts numbered I, II, and III.

There are three types of centre screw to choose from. The most usual are stainless steel screws suitable for constant feeding of almost all kinds of filling materials such as cream, jam, fish, bean paste, minced meat, or a mixture of vegetables and meat. A second type is specially used for bean paste. A third type is exclusively used for soft fillings such as cream and jam.

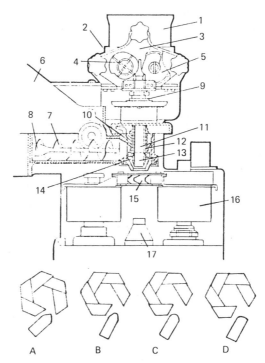

Fig. 13.4 Flourless encrusting moulding, and four types of shutter (Courtesy of Rheon Automatic Machinery Co Ltd, Japan) (1) Centre (filling) hopper, (2) Centre cylinder, (3) Filling controller, (4) Wing-filling controller, (5) Centre screw, (6) Side (casing) hopper, (7) Side screw, (8) Side cylinder, (9) Filling guide, (10) Compound nozzle, (11) Inner nozzle, (12) Outer nozzle, (13) Nozzle tip, (14) Retaining ring, (15) Shutter—A, B, C, D, (16) Flourless encrusting unit, (17) Junk cap.

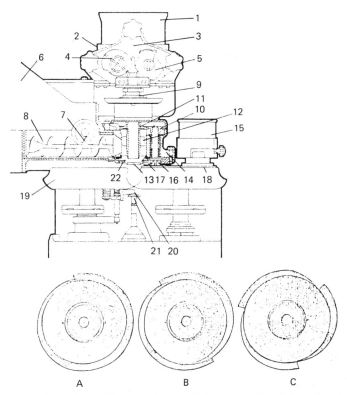

Fig. 13.5   Encrusting disk moulding and the one-sided two-thread, and three-thread disks A, B, C (Courtesy of Rheon Automatic Machinery Co. Ltd, Japan) (1) Centre (filling) hopper, (2) Centre cylinder, (3) Filling controller, (4) Wing-filling controller, (5) Centre screw, (6) Side (casing) hopper, (7) Side screw, (8) Side cylinder, (9) Filling guide, (10) Compound nozzle gear, (11) Turning nozzle holder, (12) Turning nozzle, (13) Turning ring, (14) Compound nozzle, (15) Flour hopper, (16) Flour dusting brush, (17) Flour dusting pan, (18) Flour distributing disk, (19) Encrusting disk, (20) Batting arm, (21) Stability disk, (22) Thrust washer.

13.2.2.3 Co-extrusion assembly

This assembly plays a very important role in the formation of the encrusting food by combining the casing and filling material into a tube-like product. For spherical products, the unit is made up of a turning ring which is bolted at the bottom of the compound nozzle by a thrust washer and cover, and a flour dusting brush assembled on the cover and driven by corresponding gears.

Above the turning ring is the turning nozzle which is screwed to its holder. It introduces the filling material to meet the casing material at the bottom of the compound nozzle. Here the side face is connected with the side cylinder opening by means of clamp screws.

For cylindrical products, the co-extrusion assembly is similar to the one for spherical products, and will be discussed in section 13.3.1.

13.3 MOULDING

There are three types of moulding assembly for the Rheon 208, which make either spherical or cylindrical products. Of the three, the encrusting disk assembly is suitable for the moulding of highly elastic materials such as bread doughs and hard doughs which require rounding during shaping. The flourless encrusting unit is specifically for moulding highly viscous materials such as minced meat, with which the use of dusting flour during shaping is unfavourable. The third type of moulding unit, the extrusion nozzle, is for cookie bars, filled doughnuts, sausage rolls, meat rolls, and twisted doughnuts, etc.

13.3.1 Extrusion nozzle moulding

Compared with the other two types of moulding, this moulding unit is simpler in structure and easier to understand and operate (cylindrical moulding, Fig. 13.3). It is composed of an inner nozzle and an outer nozzle which is a hollow screw to transfer the casing material downward and to allow the inner nozzle to guide the filling material as it goes down. The outer nozzle is like the turning nozzle in the spherical moulding, but under it is a fixed ring (not a turning one). When the casing and filling material from the outer nozzle and inner nozzle meet and are extruded through the fixed ring, a filled cylindrical (tube-like) product is formed. Because of this, this shaping process is also called a cylindrical moulding.

The shapes of casing and filling materials can be achieved by different combinations of the various inner nozzles and fixed ring, which make it possible to produce a wide variety of filled cylindrical products with graceful patterns from 8 mm to 60 mm in diameter. Furthermore, there are many kinds of shaping option which can be fixed on the option base at the conveyor section and synchronized with the conveyor speed to perform secondary shaping work by means of rolls, wire cutters, printers, sprayers, etc.

The extruded cylindrical product drops continuously onto the conveyor web which leads either to the secondary shaping or directly to the baking equipment.

13.3.2 Flourless encrusting moulding

This unit (see Fig. 13.4) is suitable for the highly viscous and highly elastic materials used to produce closed filled bars or balls in various sizes by flourless moulding and co-extrusion. The co-extrusion assembly consists of inner and outer nozzles, a compound nozzle, a nozzle tip fixed at the lower end of the outer nozzle, and a retaining ring. When the extruded cylindrical product passes downward, the underlying shutter will shape it into a preset size and cut it into balls. The shutter comprises six small movable blocks controlled by a set of mechanisms sliding either to enlarge or close the opening so that the filled cylinder is cut and shaped into balls.

There are four types of shutter. Type A is suitable for moulding cookies, doughnuts, cheese and potato paste. Type B is used for ground meat, fish paste, marzipan pie. Type C is for mochi, and other elastic casing materials. Type D is for mochi and other highly elastic materials.

This flourless encrusting unit can produce a wide range of products which can be closed filled bars or balls with a maximum length of 120 mm and a weight range of 40 g to 80 g.

13.3.3 Encrusting disk moulding

This moulding assembly (Fig. 13.5) consists of the co-extrusion compound nozzle unit, encrusting disk unit, batting arm, stability disk, work conveyor, and flour dusting unit.

The co-extrusion compound nozzle unit is similar to that of the other two moulding assemblies, only the inner nozzle and nozzle tip are missing, and a turning ring is mounted instead of the fixed ring used in the extrusion nozzle moulding and flourless encrusting unit.

The flour dusting unit (duster) is necessary with this moulding process. A suitable amount of dusting flour guarantees product quality. The flour hopper is fixed on its support near the right hand encrusting disk on which the dusted flour is divided by the flour distribution disk into two routes. One route runs under the compound nozzle to the dusting brush, and the other continues to the circumference of the encrusting disk, from which the flour is supplied onto the side face of the product to prevent the material sticking to the encrusting disk. The surplus flour is collected by an accessory flour rake into the flour receptacle mounted on the side of the machine for re-use.

The most advanced technique of the Rheon encrusting machine is exhibited in its unique encrusting disk moulding. Two encrusting disks are essential for rounding and cutting the products. Significant features include a special spiral thread with gradually changing major and minor diameters of the spiral, the shape and dimensions of the thread profile, and the thread angle. All these changes are in perfect harmony, so that as the height of thread engagement becomes larger, the pitch becomes greater, the thread angle smaller, and the major diameter of the coaxial cylinder becomes gradually longer. The two revolving disks make the engaged cylindrical casing material gradually migrate to the centre part of the original cylinder. When the highest points of the thread on the two counter-rotating disks tangentially meet at the centre of the cylinder product (that is, the two changing threads on their larger major diameter are tangential to each other), the cylinder is cut and closed completely into two hemispherical parts, rounded by the revolving spiral surface of the disks.

The stability disk plays the role of a bracket plate for the spherical products in its top position. As soon as the rounded piece departs from its main body (filled tube), the stability disk descends to its lower position where the product is pushed by the batting arm and received by the work conveyor leading to secondary shaping by preset options.

This moulding method is commonly called 'interaction forming process' which is directed towards force delivery. Since all the parameters of the thread (the shape and dimensions of the thread profile as well as the major and minor diameters of the spiral) manufactured on the disk side surface are gradually changing in coordination with each other, only a small part of the casing material is engaged by the counter-revolving disks and forced to move in the predetermined direction (to form a ball) at any moment. Subsequently, the filling material is separated little by little. The local motion of a small part of the casing material takes it to its destined position under a constant and continuous force gently applied by the rotating disks.

A series of location motion and partial displacement of the casing material leads to an interlocking reaction of the cylindrical product extruded by the compound nozzle assembly until the casting material migrates to the very centre and is completely separate from the filling material. By that time, the highest points of the thread on the two rotating disks tangentially meet at the centre to cut the ball off.

This encrusting moulding process is eminently suitable for producing various cakes. In this way, the compound movement of the revolving encrusting disks with the changing thread surface and the descending bar ensures that there is no constant contact between the applied force and the object surfaces. This means that the product formation is done gradually and continuously without stress concentration in the formed product which is commonly caused by the deep-hollow-punching method, and which results in undesirable compacted texture and a harder sensation.

There are three types of encrusting disk with single, double, or triple forming and cutting threads to produce one, two, or three balled products at each rotation. The product weight range is from 7 g to 250 g. The more threads on the encrusting disk, the more pieces of spherical product will be produced per hour, and the smaller the ball will be.

For the use of multi-thread encrusting disks, the cam (68) (Fig. 13.6) for the stability disk should be changed to make the stability disk move up and down at a speed synchronized with the encrusting moulding rate. The corresponding batting arm is also multi-fixed on the batting holder connected to the right-hand disk.

13.4   TRANSMISSION SYSTEM

The Rheon encrusting machine is a multi-functional unit with a rather complicated transmission system. Fig. 13.6 is a schematic diagram designed to help understanding. Some of the gear shaftings and other mechanisms have been simplified, but it admittedly remains rather complex.

From motor (1) (2.6 kW, 3-phase or 2.2 kW for SS-Type) the rotation is transmitted through the belt pulley reducers (2), (3) to shaft I which distributes the power in two ways: one is through the stepless reducer (4), intermediate shaft II, and worm-gear mechanism (5) and (6) to shaft III; the other is by means of the belt-pulley system (37) and (38), shaft XVIII, and worm-gear mechanism (39) and (40) to shaft XVII. The main spindles, shafts III and XVIII, transmit the power and motion through a series of gear trains, chain drives, gear shifts, cam gear, and intermediate shafts to the corresponding working mechanisms and systems.

(A) From shaft III, the motion is transmitted further via two branches: One is through the worm gearing 7 and 50, shaft IV, gear shifting (9), (10), (11), and (12), gear (13) to shaft VI. The other is through the worm gearing (7) and (8), shaft XI, gear wheel (24), (25), shaft XII, gearwheel (26) and (27) to shaft XIII. From shaft VI and XIII, the rotation is further transmitted in four sub-branches, two for each shaft separately.

(a) From shaft VI, shaft VIII receives the power through chain wheels (14) and (15), shaft VII, and gear wheels (16), (17) so that the side screw (18) is made to rotate to convey the casing materials to the outer spiral thread of the turning nozzle (43).

From shaft VI, shaft X receives torque through gear wheels (19) and (20), shaft IX, bevel gears (21) and (22), so that the stability roller (23) runs to keep the casing materials flowing steadily into the turning nozzle (43).

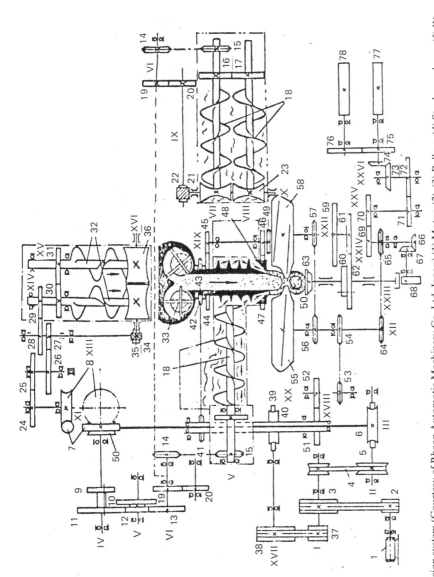

Fig. 13.6   Transmission system (Courtesy of Rheon Automatic Machinery Co. Ltd, Japan) (1) Motor, (2), (3) Pulleys, (4) Stepless reducer, (5) Worm, (6) Worm gear, (7), Worm, (8) Worm gear, (9)–(13) Gears, (14), (15) Chain wheels, (16), (17) Gears, (18) Side screw, (19)–(22) Gears, (23) Stability roller wing, (24)–(31) Gears, (32) Centre screw, (33) Filling controller, (34), (35) Gears, (36) Filling stability roller, (37), (38) Pulleys, (39) Worm, (40) Worm gear, (41), (42) Chain wheels, (43) Turning nozzle, (44)–(47) Gears, (48) Turning ring, (49) Gear, (50) Product, (51), (52) Gears, (53), (54) Chain gears, (55) Left encrusting disk, (56), (57) Chain wheels, (58) Right encrusting disk, (59)–(62) Gears, (63) Stability disk, (64), (65) Chain wheels, (66), (67) Bevel gears, (68) Cam, (69)–(76) Gears, (77), (78) Conveyor roller. I–XXVI Shafts.

(b) From shaft XIII, the rotation is transmitted through gear wheels (28) and (29), shaft XIV, gear wheels (30) and (31), to shaft XV so that the centre screw (32) and filling controller wing (33) push the filling materials down into the turning nozzle (43) through the filling guide.

From shaft XIII, the power is transmitted through bevel gear wheels (34) and (35) to the filling stability roller (36) which gives a stable flow to the filling materials.

(B) From shaft XVIII, the rotation is transmitted through the chain gearing (41) and (42) to the turning nozzle (43) which is actually a hollow screw allowing the transfer of the casing material by its outer thread and permitting the filling material to go down the inside. The casing and filling material meet at the turning ring (48).

Also from the chain wheels 41 and 42, the rotation is transmitted through the gear wheels (44) and (45), intermediate shaft XIX, and gear wheels (46) and (47) to the turning ring (48) which receives the casing and filling materials and shapes them into a sausage-like tube to a preset diameter and with the required filling to casing ratio. By the rotation of the ring the shaping force is gently applied to the food material so as to cause no harm to the texture and actual sensation of the finished product.

At the same time, shaft XIX also delivers motion through the gear (49) to the dusting brush which distributes the flour to the product surface and the side thread of the encrusting disk to prevent the casing material sticking to the disk during shaping.

From shaft XVIII, rotation is transmitted through the gear wheels (51) and (52), shaft XX, and chain wheels (53) and (54) to shaft XXI which delivers the power and motion in three branches. One is directly to the left encrusting disk (55) which cooperates with the right encrusting disk (58) to shape and cut the casing and filling materials into balls. The second is through chain wheels (56) and (57) to shaft XXII which transmits rotation in two sub-branches. The last one is through chain wheels (64) and (65) to shaft XXIV, which drives the stability disk and conveyor by three intermediate shafts and a series of mechanisms.

$$\text{Motor 1} \rightarrow \frac{\text{Pulley 2}}{\text{Pulley 3}} \rightarrow \text{Shaft I} \begin{cases} \text{Stepless reducer 4} \rightarrow \text{Shaft II} \\[2mm] \dfrac{\text{Pulley 37}}{\text{Pulley 38}} \rightarrow \text{Shaft XVII} \end{cases}$$

$$\text{Shaft II} \rightarrow \frac{\text{Worm 5}}{\text{Worm gear 6}} \rightarrow \text{Shaft III} \rightarrow \frac{\text{Worm 7}}{\text{Worm gear 50}} \rightarrow \text{Shaft IV}$$
$$\hookrightarrow \frac{\text{Worm 7}}{\text{Worm gear 8}} \rightarrow \text{Shaft XI}$$

$$\text{Shaft IV} \begin{cases} \text{Gear 9} \\ \overline{\text{Gear 10}} \\ \text{Gear 11} \\ \overline{\text{Gear 12}} \end{cases} \text{Gear 13} \rightarrow \text{Shaft VI} \rightarrow \frac{\text{Chain wheel 14}}{\text{Chain wheel 15}} \rightarrow \text{Shaft VII}$$
$$\hookrightarrow \frac{\text{Gear 19}}{\text{Gear 20}} \rightarrow \text{Shaft IX}$$

Shaft VII $\to \dfrac{\text{Gear 16}}{\text{Gear 17}} \to$ Shaft VIII $\to$ Side screws 18 (Dough feeding)

Shaft IX $\to \dfrac{\text{Bevel gear 21}}{\text{Bevel gear 22}} \to$ Shaft X $\to$ Stability roller 23

Shaft XI $\to \dfrac{\text{Gear 24}}{\text{Gear 25}} \to$ Shaft XII $\to \dfrac{\text{Gear 26}}{\text{Gear 27}} \to$ Shaft XIII

Shaft XIII $\begin{cases} \dfrac{\text{Gear 28}}{\text{Gear 29}} \to \text{Shaft XIV} \to \dfrac{\text{Gear 30}}{\text{Gear 31}} \to \text{Shaft XV} \begin{cases} \text{Centre screw 32} \\ \text{Filling} \\ \text{controller wing 33} \end{cases} \\[4ex] \dfrac{\text{Bevel gear 34}}{\text{Bevel gear 35}} \to \text{Shaft XVI} \to \text{Filling stability roller 36} \end{cases}$

Shaft XVII $\to \dfrac{\text{Worm 39}}{\text{Worm gear 40}} \to$ Shaft XVIII.

Shaft XVIII $\begin{cases} \dfrac{\text{Chain wheel 41}}{\text{Chain wheel 42}} \to \begin{cases} \to \text{Turning nozzle 43} \\[2ex] \dfrac{\text{Gear 44}}{\text{Gear 45}} \to \text{Shaft XIX} \begin{cases} \dfrac{\text{Gear 46}}{\text{Gear 47}} \to \text{Turning ring 48} \\[2ex] \text{Gear 49} \to \text{Dusting brush} \end{cases} \end{cases} \\[6ex] \dfrac{\text{Gear 51}}{\text{Gear 52}} \to \text{Shaft XX} \to \dfrac{\text{Chain wheel 53}}{\text{Chain wheel 54}} \to \text{Shaft XXI} \end{cases}$

Shaft XXI $\begin{cases} \dfrac{\text{Chain wheel 56}}{\text{Chain wheel 57}} \to \text{Shaft XXII} \to \begin{cases} \text{Left encrusting disk 55} \\[1ex] \text{Right encrusting disk 58} \\[1ex] \begin{cases} \text{Gear 59} \\ \dfrac{\text{Gear 60}}{\text{Gear 61}} \\ \text{Gear 62} \end{cases} \text{Shaft XXIII} \to \begin{matrix} \text{Stability} \\ \text{disk 63} \\ \text{(revolving)} \end{matrix} \end{cases} \\[10ex] \dfrac{\text{Chain wheel 64}}{\text{Chain wheel 65}} \to \text{Shaft XXIV} \begin{cases} \dfrac{\text{Bevel gear 66}}{\text{Bevel gear 67}} \to \text{Cam 68} \to \begin{matrix} \text{Stability} \\ \text{disk 63} \\ \text{(up \& down)} \end{matrix} \\[3ex] \dfrac{\text{Gear 69}}{\text{Gear 70}} \to \text{Shaft XXV} \end{cases} \end{cases}$

Shaft XXV $\to \dfrac{\text{Gear 71}}{\text{Gear 72}} \to$ Shaft XXVI $\to \dfrac{\text{Bevel gear 73}}{\text{Bevel gear 74}} \to \begin{cases} \text{Conveyor roll 77} \\[2ex] \dfrac{\text{Gear 75}}{\text{Gear 76}} \to \text{Conveyor roll 78} \end{cases}$

Scheme 13.1.   Transmission system of an encrusting machine.

From shaft XXII, rotation is directly transmitted to the right encrusting disk (58) to cooperate with left encrusting disk (55).

Shaft XXIII, actuated by the gear shifts (59), (60), (61), and (62) on shaft XXII, drives the stability disk to match the rotation of the encrusting disks to provide a synchronized shaping performance.

From shaft XXIV, the rotary action is changed into an up-and-down reciprocating action to the stability disk (63) by means of the bevel gears (66) and (67) and cam mechanism (68). The up-and-down motion of the stability disks (63) matches the product shaping speed, moving up to receive the product, and then going down with the product. By the time it arrives at the shaping end, the rotating batting arm also arrives and pushes the product from the stability disk onto the work conveyor.

Also from shaft XXIV, the rotation is transmitted through intermediate shafts XXV and XXVI, gear trains 71 and 72, 73 and 74, 75 and 76 to the conveyor rollers 77 and 78 which drive the endless web to deliver the products to the next stage, either for secondary shaping or directly to the cooking equipment. Scheme 13.1 is the transmission expression of the encrusting machine.

13.5   OPERATION

This machine needs more attention and its operators more instruction before starting, as its desired performance is dependent upon a series of mechanisms operating in perfect harmony.

Before starting the machine, the operator should check if the filling controller is correctly fixed, whether the compound nozzle is in the proper operating position, and so on. Correct adjustment of the appropriate controls allows many numerous different food products to be produced.

13.5.1   Filling control

Two steps are used to vary the filling output by means of a centre clutch. Step (A) allows a greater output of filling. That is, if the step is changed from (A) to (B), the filling output is reduced. Four choices of ratio (A to B) are available: 1:0.2, 1:0.3, 1:0.5, and 1:0.65. Of these 1:0.5 is normally used. However, the filling output is affected by the state of the filling material in the centre hopper. If it is pressed by hand, it will become rigid and tend to bridge above the two centre screws inside the hopper and fail to be engaged by the screw flukes. This results in irregular output.

13.5.2   Casing thickness control

The casing output can be adjusted by means of the side clutch. There are five steps numbered 1, 2, 3, 4 and 5 in order from a small output, while the rotation ratio of the side screws at each step is shown on the specification plate near this side clutch. In fact, the thickness of the casing is dependent upon the filling to casing ratio. That is, the side and centre clutches should be adjusted in cooperation to obtain ten types of filling to casing ratio. Of course, the way that the side hopper is loaded is also important for stable output and correct filling to casing ratio. The casing materials should not cover the stability roller, otherwise more or less air will form pockets above the side screws inside the hopper. This occurs easily when using yeast-raised or pastry doughs. If this happens, uncover the stability roller and slightly press the dough to eliminate the air pocket by hand without touching the revolving screws, even though they are working at a very low speed. The physical properties of different materials used may vary in viscosity, density, water absorption, amount of fat, salt or syrup content, type of wheat and size of filling pieces which will not always permit the output to be altered in proportion to the variation of the screw rotation.

Table 13.2 shows the filling to casing ratio and the weight range of products at five different steps for A and B when a 30 rotation-type is used for buns. The figures are optimum in terms of product finish and stability. The '30 rotation-type' produces 30 pieces per minute.

13.5.3   Product size adjustment

The final size of the product is controlled by turning the size adjustment handle without changing the filling to casing ratio. Clockwise rotation makes the product smaller, while counterclockwise rotation leads to a larger size. It is necessary to remember that any change should be carried out only when the machine is not running. The handle must be pulled out before turning since it is spring loaded, which allows the handle to be locked in position after being released. The 'large' and 'small' as well as a double ended arrow signal are marked on the handle near to which a size indicating meter is fixed. Since the product weight varies, depending not only on the clutch step selected but also on the specific gravity of the casing and filling materials used the meter indication is calibrated in arbitrary units, from which a record of the batching position and the clutch step used and the actual product weight can be easily

Table 13.2   Filling to casing ratio and the weight range of the encrusted products
(30 rotation-type)

Turning nozzle Turning ring		For doughnuts No. 19			For bun (No. 80) No. 18		
Clutch	Ratio/Weight	Ratio Filling:Casing	Min (g)	Max (g)	Ratio Filling:Casing	Min (g)	Max (g)
A	1	1:0.45	44	100	1:0.49	45	100
	2	1:0.61	50	105	1:0.58	51	105
	3	1:0.86	55	110	1:0.87	56	110
	4	1:1.01	58	120	1:1.03	59	120
	5	1:1.25	58	120	1:1.44	69	120
B	1	1:0.59	33	78	1:0.75	34	78
	2	1:0.87	38	87	1:1.01	39	89
	3	1:1.19	43	100	1:1.25	44	100
	4	1:1.53	48	110	1:1.50	49	110
	5	1:1.88	52	120	1:1.87	53	120

calibrated. Care must be taken that the red zones at either end of the meter scale are avoided. They indicate serious overload. The machine cannot be used continuously in an overloaded condition.

Depending on the machine type, the size adjusting handle allows a wide weight range of 10 g–40 g, 20 g–240 g, 25 g–80 g, 30 g–100 g, 30 g–120 g, 35 g–150 g and 40 g–100 g.

13.5.4   Flour dusting regulation
Insufficient flour dusting will cause the materials to stick to the encrusting disks, while an excessive supply will lead to irregular cut-ends on the top of the products. Flour dusting is controlled by the regulating knob on the flour unit. Turning it clockwise increases the flour flow, and counterclockwise decreases it.

13.5.5   Stability disk adjustment
The height and size of the stability disk are very important for the shape and quality of the product. They should be selected for the correct size and placed in the best position to suit the weight of the product. The adjustment handle is fixed at the lower part of the main body. Turning the handle clockwise raises the disk, and counterclockwise lowers it. As soon as the handle meets resistance, it should not be turned further, since the adjustment limit has been reached. If the stability disk needs to be raised further, the disk elevating locknut on the sliding shaft should be employed. If the product is flat on top and thicker at the sides, it means that the disk is set too high. If the product is poorly shaped with an irregular loading, it results from too low a position of the disk. It is also necessary to adjust the clearance between the stability disk and the batting arm in the range of 1 mm to 2 mm after each disk height alteration.

NOTE All diagrams and data in this chapter by courtesy of Rheon Automatic Machinery Co. Ltd, Japan.

Part VII

Baking ovens

14

Heat transfer principles and electric heating elements

Baking ovens perform the vital function of converting the unpalatable raw dough into ready-to-eat food through the medium of heat.

Heat is a form of energy, and the temperature is a measurement scale for thermal energy.

The first law of thermodynamics states, 'Energy cannot be created or destroyed, it is merely transformed from one form to another.' The central theme of the second law of thermodynamics is 'Energy will flow spontaneously only from a region having high energy to another that is at a lower energy level.' That is, the heat exchange existing between any objects with different temperatures. So when the dough pieces at the low temperature of the factory are conveyed into the high temperature oven chamber, there must be heat exchange.

In baking ovens, heat is transferred by a combination mode of three different mechanisms: conduction, convection, and radiation. We shall review briefly the principles of heat transfer before discussing heating elements and oven constructions.

14.1 HEATING TRANSFER PRINCIPLES

14.1.1 Conduction

Conduction is the transfer of heat which takes place at a molecular level. When molecules of a solid object have thermal energy, they vibrate at their respective locations: the heat, thermal energy, is transferred through these vibrations from one molecule to another adjacent to it without actual transmission of the molecules through the intervening space. This is the primary heat transfer mechanism in solids.

If a temperature gradient exists in a body or between two bodies in physical contact, heat conduction will take place from the high temperature region to the low temperature region. The fundamental differential equation for heat conduction is called Fourier's law:

$$Q_x = -kA \frac{\mathrm{d}t}{\mathrm{d}x} \tag{14.1}$$

where Q_x is the heat flow rate in direction x by conduction (W), A is the area normal to direction x in which the heat flows (m²), dt/dx is the rate of temperature change (°C) with the distance (m) in direction x of the heat flow, that is, the temperature gradient, t is the temperature, x is the distance measured along the direction x of the heat flow. The negative sign indicates that heat always flows from the higher temperature region, which is in accordance with the second law of thermodynamics, k is the thermal conductivity of a product in quality terms, that is the rate of heat conduction through a unit thickness of the material, as a unit temperature gradient exists across that thickness. In SI units, $k = $ W/m.°C.

Factor k is dependent upon temperature and upon the material through which the heat flows. Table 14.1 gives the thermal conductivity of various commonly encountered materials. The thermal conductivity of selected food products is given in Table 14.2.

The materials with high heat conductivity such as steel can be used as dough piece carriers (trays or bands) for raising the heat transfer efficiency, since in tray or band ovens the heat transfer between the dough piece and the carrier is by typical heat conduction.

The materials with low thermal conductivities such as glass reinforced plastic and mineral wool are often used as insulating materials in the oven chamber walls to reduce heat losses through the walls by conduction.

The thermal conductivities of various high moisture dough pieces on entering the oven have values close to the thermal conductivity of water (0.597 W/m.°C at 20°C). As the bottom of a dough piece is heated by the high temperature oven band or tray through heat conduction, steep temperature gradients are set up within the dough piece, which result in vigorous vibration of molecules and movement of liquids within the dough. The vibration causes heat conduction, while the liquid movement causes heat convection.

Table 14.1   Thermal conductivity of various materials†

	$\dfrac{\text{BTU}}{\text{h(ft) (°F)}}$	$\dfrac{\text{W}}{\text{m(°C)}}$
Building materials:		
Building brick	0.40	0.69
Building plaster	0.25	0.43
Concrete	0.54	0.93
Concrete blocks		
two oval core, 8 in thick	0.60	1.04
two rectangular core, 8 in thick	0.64	1.11
Corkboard	0.025	0.043
Felt (wool)	0.03	0.052
Glass	0.3–0.61	0.52–1.06
Gypsum or plasterboard	0.33	0.57
Wood (laminated board)	0.045	0.078

Table 14.1 (*continued*)

	BTU h(ft) (°F)	W m(°C)
Wood (across grain, dry)		
maple	0.11	0.19
oak	0.12	0.21
pine	0.087	0.15
Wood (plywood)	0.067	0.12
Rubber (hard)	0.087	0.15
Insulating materials:		
Air		
32°F(0°C)	0.014	0.024
212°F(100°C)	0.0183	0.032
392°F(200°C)	0.0226	0.039
Fibreglass (9 lb/ft^3 density)	0.02	0.035
Polystyrene		
2.4 lb/ft^3 density	0.019	0.032
2.9 lb/ft^3 density	0.015	0.026
1.6 lb/ft^3 density	0.023	0.040
Polyurethane (5–8.5 lb/ft^3 density)	0.019	0.033
Hog hair with asphalt binder		
(8.5 lb/ft^3 density)	0.028	0.048
Mineral wool with binder	0.025	0.043
Metals:		
Aluminium		
32°F(0°C)	117	202
212°F(100°C)	119	205
572°F(300°C)	133	230
Cast iron		
32°F(0°C)	32	55
212°F(100°C)	30	52
572°F(300°C)	26	45
Copper		
32°F(0°C)	294	509
212°F(100°C)	218	377
572°F(300°C)	212	367
Steel (carbon)		
212°F (100°C)	26	45
572°F (300°C)	25	43
Steel, stainless Type 304 or 302	10	17
Steel, stainless Type 316	9	15

†Adapted from Romeo T. Tolendo (1980) Copyright © (1980) by Van Nostrand Reinhold. (AVI Publishing Company, Inc. USA). Reprinted by permission of the publisher

Table 14.2   Thermal conductivity of selected food products[a]

Product	Moisture content (%)	Temperature (°C)	Thermal conductivity (W/m · °C)
Apple	85.6	2–36	0.393
Apple sauce	78.8	2–36	0.516
Beef, freeze dried			
1000 mm Hg pressure	—	0	0.065
0.001 mm Hg pressure	—	0	0.037
Beef, lean			
Perpendicular to fibres	78.9	7	0.476
Perpendicular to fibres	78.9	62	0.485
Parallel to fibres	78.7	8	0.431
Parallel to fibres	78.7	61	0.447
Beef fat	—	24–38	0.19
Butter	15	46	0.197
Cod	83	2.8	0.544
Corn, yellow dust	0.91	8–52	0.141
(Maize)	30.2	8–52	0.172
Egg, frozen whole	—	− 10 to − 6	0.97
Egg, white	—	36	0.577
Egg, yolk	—	33	0.338
Fish muscle	—	0–10	0.557
Grapefruit, whole	—	30	0.45
Honey	12.6	2	0.502
	80	2	0.344
	14.8	69	0.623
	80	69	0.415
Juice, apple	87.4	20	0.559
	87.4	80	0.632
	36.0	20	0.389
	36.0	80	0.436
Lamb			
Perpendicular to fibre	71.8	5	0.45
		61	0.478
Parallel to fibre	71.0	5	0.415
		61	0.422
Milk	—	37	0.530
Milk, condensed	90	24	0.571
	—	78	0.641
	50	26	0.329
	—	78	0.364

Table 14.2 (*continued*)

Product	Moisture content (%)	Temperature (°C)	Thermal conductivity (W/m · °C)
Milk, skimmed	—	1.5	0.538
	—	80	0.635
Milk, nonfat dry	4.2	39	0.419
Olive oil	—	15	0.189
	—	100	0.163
Oranges, combined	—	30	0.431
Peas, black-eyed	—	3–17	0.312
Pork			
Perpendicular to fibres	75.1	6	0.488
		60	0.54
Parallel to fibres	75.9	4	0.443
		61	0.489
Pork fat	—	25	0.152
Potato, raw flesh	81.5	1–32	0.554
Potato, starch gel	—	1–67	0.04
Poultry, broiler muscle	69.1–74.9	4–27	0.412
Salmon			
Perpendicular	73	4	0.502
Salt	—	87	0.247
Sausage mixture	65.72	24	0.407
Soybean oil meal	13.2	7–10	0.069
Strawberries	—	14 to 25	0.673
Sugars	—	29–62	0.087–0.22
Turkey, breast			
Perpendicular to fibres	74	3	0.502
Parallel to fibres	74	3	0.523
Veal			
Perpendicular to fibres	75	6	0.476
		62	0.489
Parallel to fibres	75	5	0.441
		60	0.452
Vegetable and animal oils	—	4–187	0.169
Wheat flour	8.8	43	0.45
		65.5	0.689
		1.7	0.542
Whey		80	0.641

14.1.2   Convection

This is the transmission of heat from one part to another, whenever there is a temperature difference between the two, through a medium capable of free circulation such as gas, steam, or liquid. Different from conduction, heat convection is achieved by the flow of a fluid in contact with the object to be heated, that is, by the relative movement between liquid and solid, or liquid and liquid. In an oven, molecules of high temperature air, steam, or combustion gases, circulate throughout the baking chamber and transfer heat to the surfaces of the dough pieces as they are in contact with each other. Since the temperatures of the two are different, molecules of the hot fluid would continuously collide with the surface of the low-temperature dough pieces, exchange energy in the collision process, and eventually leave the heated dough piece surfaces to mix with the bulk of the hot fluid. At the same time, steep differentials of temperature are set up within the dough piece, which result in movement of water and its vapour, melted shortenings, gases, and other fluids. The movement leads to a transfer of heat from one part to another, that is, heat convection within the dough pieces.

Heat transfer by convection is expressed by Newton's law of cooling:

$$Q = h_c \, A(T_b - T_f) \tag{14.2}$$

where Q is the rate of heat transfer (W), h_c is the convective heat transfer coefficient (also referred to as the surface heat transfer coefficient) (W/m^2°C), A is the area of a surface in contact (m^2), $(T_b - T_f)$ is the overall temperature gradient, T_b is the bulk temperature of the fluid in contact with the surface (°C), and T_f is the temperature of the surface (°C).

Equation (14.2) shows that heat transfer by convection is proportional to the temperature difference, $(T_b - T_f)$ and that a high value of h_c reflects a high rate of heat transfer by convection.

If the convection depends only on the natural air currents set up by differences in temperature between adjacent air masses, this convection is referred to as free convection or passive convection while forced convection is achieved when a current of air is created by a fan or blower.

The magnitude of the heat transfer coefficient h_c is dependent on the system, the geometry, the properties of the fluid, and the type of convection, natural or forced. Table 14.3 shows values of h_c for some fluids, which indicate that forced convection gives a higher value of h_c than free convection. To get a more even and rapid effect of heat distribution, the forced convection technique has been developed in many modern ovens. Fig. 14.1 shows typical forced convection in a baking chamber in which the chamber air is blown by a specially designed double inlet fan over the heat exchanger to pick up heat. The heated air is then diverted to the top and bottom ducts, situated above and below the conveyor band, by means of dampers. From these ducts the high temperature air blows onto the dough pieces. The lower temperature dough is heated up, while the hot air is cooled down by heat convection. The cooled air returns to the fan for recirculation via the top duct which is specially constructed with a central feed spine and lateral blowing ribs to allow return between the ribs. The bottom duct is a full plenum type allowing easy cleaning, and is narrower than the

Table 14.3   Some approximate values of convective heat-transfer coefficient†

Fluid	Convective heat-transfer coefficient (W/m² · °C)
Air	
Free convection	5–25
Forced convection	10–200
Water	
Free convection	20–100
Forced convection	50–10,000
Boiling water	3,000–100,000
Condensing water vapour	5,000–100,000

†Adapted from R. Paul Singh and Dennis R. Heldman 1984. Copyright © (1984) by Academic Press, Inc. Reprinted by permission of the publisher.

baking chamber to allow some return air to pass beneath it to the recirculation fan (APV Baker, UK).

An optimum forced convection would considerably reduce the oven preheating time, and greatly increase heat transfer to the products, resulting in shorter baking time and fuel economy.

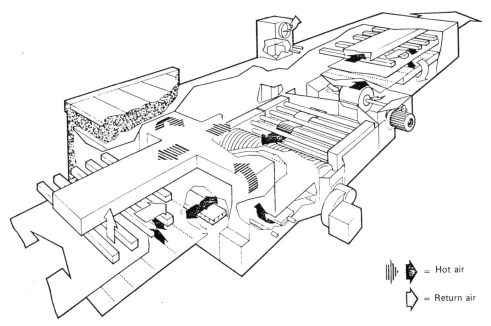

‖‖‣ ▶	= Hot air
▷	= Return air

Fig. 14.1   Forced convection in a baking chamber (Courtesy of APV Baker, UK).

14.1.3   Radiation

Radiation differs from conduction and convection. It is the heat transfer directly by electromagnetic waves and does not require any contact between objects or any transfer medium. All bodies at temperatures higher than absolute zero emit energy. It can be transmitted even in a high vacuum, and between two bodies far apart.

Energy Q emitted from surface A is in direct proportion to the nature of the surface and the fourth power of the absolute temperature. The heat flux from a surface by radiation is expressed by the following equation

$$\frac{Q}{A} = \sigma e T^4 \tag{14.3}$$

where σ is the Stefan-Boltzmann constant, equal to 5.669×10^{-8} W/m^{2}K^4, e is the emissivity describing the extent to which a surface is similar to an ideal black body for which the value of emissivity is 1 and T is the absolute temperature. Table 14.4 gives

Table 14.4   Emissivity of various surfaces[a]

Material	Wavelength and average temperatures				
	9.3 μm, 38°C	5.4 μm, 260°C	3.6 μm, 540°C	1.8 μm, 1370°C	0.6 μm, Solar
Metals					
Aluminum					
Polished	0.04	0.05	0.08	0.19	~ 0.3
Oxidized	0.11	0.12	0.18		
24-ST weathered	0.4	0.32	0.27		
Surface roofing	0.22				
Anodized (at 100°F)	0.94	0.42	0.60	0.34	
Brass					
Polished	0.10	0.10			
Oxidized	0.61				
Chromium					
Polished	0.08	0.17	0.26	0.40	0.49
Copper					
Polished	0.04	0.05	0.18	0.17	
Oxidized	0.87	0.83	0.77		
Iron					
Polished	0.06	0.08	0.13	0.25	0.45
Cast, oxidized	0.63	0.66	0.76		
Galvanized, new	0.23	—	—	0.42	0.66
Galvanized, dirty	0.28	—	—	0.90	0.89
Steel plate, rough	0.94	0.97	0.98		
Oxide	0.96	—	0.85	—	0.74
Magnesium	0.07	0.13	0.18	0.24	0.30
Silver					
Polished	0.01	0.02	0.03	—	0.11

Table 14.4 (*continued*)

Material	Wavelength and average temperatures				
	9.3 μm, 38°C	5.4 μm, 260°C	3.6 μm, 540°C	1.8 μm, 1370°C	0.6 μm, Solar
Stainless steel					
18–8, polished	0.15	0.18	0.22		
18–8, weathered	0.85	0.85	0.85		
Steel tube					
Oxidized	—	0.80			
Tungsten filament	0.03	—	—	~ 0.18	0.35[b]
Zinc					
Polished	0.02	0.03	0.04	0.06	0.46
Galvanized sheet	~ 0.25				
Building and insulating materials					
Asphalt	0.93	—	0.9	—	0.93
Brick					
Red	0.93	—	—	—	0.7
Fire clay	0.9	—	~ 0.7	~ 0.75	
Silica	0.9	—	~ 0.75	0.84	
Magnesite refractory	0.9	—	—	~ 0.4	
Enamel, white	0.9				
Paper, white	0.95	—	0.82	0.25	0.28
Plaster	0.91				
Roofing board	0.93				
Enameled steel, white	—	—	—	0.65	0.47
Paints					
Aluminized lacquer	0.65	0.65			
Lacquer, black	0.96	0.98			
Lampblack paint	0.96	0.97	—	0.97	0.97
Red paint	0.96	—	—	—	0.74
Yellow paint	0.95	—	0.5	—	0.30
Oil paints (all colours)	~ 0.94	~ 0.9			
White (ZnO)	0.95	—	0.91	—	0.18
Miscellaneous					
Ice	~ 0.97[c]				
Water	~ 0.96				
Carbon					
T-carbon, 0.9% ash	0.82	0.80	0.79		
Wood	~ 0.93				
Glass	0.90	—	—	—	(Low)

[a]Adapted from Kreith (1973). Copyright © (1973) by Harper Collins (UK). Reprinted by permission of the publisher.
[b]At 3315°C.
[c]At 0°C.

several emissivity values of some common material surfaces. The emissivity e is dependent on the characteristics of the surface. At temperature 0 K the emissivity ceases. Above 0 K, all materials emit radiation as electromagnetic waves. The higher the temperature is, the shorter the wavelength will be.

Table 14.4 indicates that the emissivity value of a polished metal surface is lower than that of its oxidized surface. Because of this, the radiators are commonly made with nonpolished surfaces. Their heat efficiency will increase with surface oxidation while Table 14.4 also shows that silicon carbide has a high emissivity value. That is why some electric ovens use silicon carbide tubes or plates as radiators in the baking chamber.

When radiation of a given wavelength is incident on an object (Fig. 14.2),

$$\phi + X + \psi = 1 \tag{14.4}$$

where ϕ is the absorptivity, X is the reflectivity, and ψ is the transmission.

Suppose there is an ideal object, a black body, and its absorptivity value is 1.0. Then, compared with the black body, all other material absorptivity values must be less than 1.0 since more or less radiation energy would be reflected and transmitted. For example, many materials are opaque to waves such as light, but they are transparent to some others such as radio waves.

The absorption of radiation energy results in an increase of temperature. Any objects with a temperature higher than 0 K emit energy. The radiated energy is different from the reflected energy. The radiation is electromagnetic, and the nature of materials may reflect some of the incident radiation, depending upon the surface absorptivity value (see equation (14.4)).

Kirchhoff's law indicates that the emissivity of a body is equal to its absorptivity for the same wavelength, that is

$$e = \phi \tag{14.5}$$

This equation means that the higher the absorptivity of a body is, the higher its emissivity will be. That is, a good absorber of radiation must also be a good radiator. Thus, to raise radiation efficiency is to make a radiator with high emissivity equal to its absorptivity. The closer to the black body the material is, the higher its absorptivity

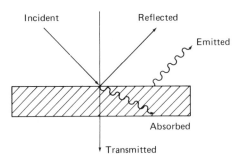

Fig. 14.2   Radiation energy incident on a semi-opaque slab.

Table 14.5 Reflectivity values of some metal surfaces [39]

Metal	Gold	Silver	Magnesium		Polished aluminium
Reflectivity	0.90	0.97	0.95		0.86
Metal	Chromium plating	Nickel plating	Polished steel plate	Zinc plating iron sheet	Aluminium foil
Reflectivity	0.72	0.72	0.54	0.65	0.98

(also its emissivity) will be. That is why new steel oven bands or trays are not as good as the old darkened ones for heat efficiency.

To increase reflectivity the oven sidewalls must reflect more energy back to the dough pieces rather than absorb it. In this way, heat losses through the walls are reduced. Modern ovens are often equipped with reflection devices in the baking chamber, and these are commonly made of polished aluminium or stainless steel with reflectivity values above 0.86. Values of reflectivity for several metals are given in Table 14.5.

14.1.4 Summary

In practice, all three types of heat transfer (conduction, convection, and radiation) are present in a baking oven chamber. As Dr S. A. Matz pointed out 'The conduction and radiation tend to cause localized temperature differentials, conduction acting to raise the temperature of the bottoms and radiation acting to raise the temperature of exposed surfaces (and especially the darkened area of protuberances), while convection tends to even out temperature gradients within the oven [8].

To raise the heat efficiency and to give full play to the advantages of convection, a convectoradiant system has been developed to provide both radiant and forced convection, while conduction takes place during baking as the dough pieces are in contact with the oven band. Fig. 14.3 illustrates this kind of oven, in which a fan

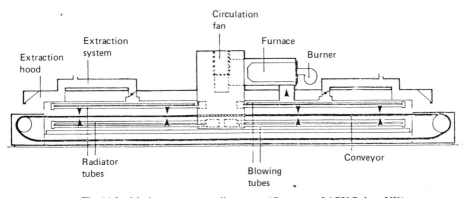

Fig. 14.3   Modern convectoradiant oven (Courtesy of APV Baker, UK).

Table 14.6   Specifications of some convectoradiant ovens (Courtesy of APV Baker, UK)

Oven size (m)	Heater units	Sea level circulating fan motors (kW)		Oven drive motor (kW)	Steam kg/h	
		1	2		Bread	Cake
10 × 2.5	1	11.1		1.2	175	23
12.5 × 2.5	1	15.0		1.2	175	23
15 × 2.5	1	15.0		1.2	175	23
17.5 × 2.5	2	15.0	4.1	2.9	262	34
20 × 2.5	2	11.1	11.1	2.9	262	46
22.5 × 2.5	2	11.1	15.0	2.9	262	46
25 × 2.5	2	11.1	15.0	2.9	349	46
27.5 × 2.5	2	15.0	15.0	2.9	349	46
30 × 2.5	2	15.0	15.0	2.9	349	46
10 × 1.65	1	4.1		1.2	113	23
12.5 × 1.65	1	5.6		1.2	113	23
15 × 1.65	1	11.1		1.2	113	23
17.5 × 1.65	2	5.6	4.1	2.9	169	34
20 × 1.65	2	4.1	4.1	2.9	169	46
22.5 × 1.65	2	4.1	4.1	2.9	169	46
25 × 1.65	2	4.1	5.6	2.9	226	46
27.5 × 1.65	2	5.6	11.1	2.9	226	46
30 × 1.65	2	11.1	11.1	2.9	226	46

Extract fan motor 1.2 kW throughout range

installed on the top of the oven circulates the hot gases produced by the burner through radiation tubes above and below the conveyor, thence into blowing tubes, which run parallel to, and in between the radiator tubes. Finally, they are discharged into the baking chamber to continue the cycle of recirculation. Thus, the system evens out the temperature gradients caused by radiation, and greatly increases heat transfer to products by convection, resulting in shorter oven preheating and baking time and fuel economy. Table 14.6 shows specifications of convectoradiant ovens.

14.2   OVEN CLASSIFICATIONS

Early oven chambers were constructed of brick or stone, and were heated directly by maintaining a wood fire within its hearth until the oven had acquired and stored sufficient heat to complete a baking run. The fire then had to be raked out from the baking hearth to start the baking operation. Peels (long-handled shovels) were used for loading and unloading this kind of oven. Because of this, the ovens were consequently referred to as peel brick ovens [2].

Modern techniques ensure that many kinds of advanced oven have been developed and have replaced the old peel brick ovens from which almost all modern ovens are derived.

There are several possible ways to classify ovens.

According to the type of energy sources, they may be oil ovens, gas ovens, fuel ovens, electric ovens, and steam ovens.

According to the heating system, they may be direct fired or indirect fired ovens, hydrid ovens, and combination ovens.

According to the type of dough piece carrier, they may be tray ovens or conveyor ovens. In the former group, there are travelling tray ovens including travelling-tray-tunnel ovens and travelling-tray-reel ovens and stationary tray ovens with shelf doors; In the latter group, there are solid-steel-sheet-band ovens, perforated-band ovens, and wire-mesh-band ovens with continental wire mesh, heavy mesh, and weave mesh respectively.

According to the form of oven chamber, they are classified into tunnel ovens, single lap and double lap ovens, reel ovens, and rack ovens.

Electric ovens, although not new, deserve some special consideration because it is probable that they will assume a much more important role in the near future when the combustion of convenient hydrocarbons such as oil and gas become much more expensive [11]. Tunnel electric ovens play a more and more important role in today's bakery industry, for they are simple in construction, and extremely convenient to manufacture, install, control, and maintain.

14.3 ELECTRIC HEATING ELEMENTS—RADIATORS

Thermal radiation is defined as radiant energy emitted by a medium that is due solely to the temperature of the medium; that is, it is the temperature of the medium which governs the emission of thermal radiation [19]. The wavelength of thermal radiation is approximately 0.3 μm to 50 μm. This wavelength has three sub-ranges, ultraviolet, visible, and infrared. The classification of thermal radiation wavelengths is given in Fig. 14.4. In bakery ovens, the commonly used wavelength is 2.5 μm to 15 μm [21].

In electric ovens, heat radiation performs the most important role in the baking process. Electric heating elements are used as radiation media, that is, radiators by which the baking chamber is heated up quickly. When the sidewalls, oven band, chamber roof and bottom, and other parts inside the oven are heated up sufficiently,

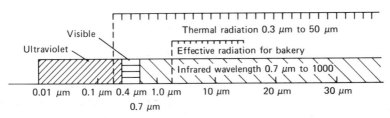

Fig. 14.4   Classification of radiation wavelength. Wavelengths in micrometres (μm).

they become in turn sources of heat radiation themselves and give up their thermal energy to the baking products.

The electric heating elements in food ovens are commonly designed in tubes or slabs with the electric heater elements inside.

14.3.1   Tube radiators

This type of radiator includes the metal duct radiator, the silicon carbide duct radiator, the silicon carbide rod radiator, and the opal quartz duct radiator, which is a newly developed type and is not used as widely as the other three, which are the most popular baking ovens.

14.3.1.1   Magnesium oxide-metal duct radiators

As shown in Fig. 14.5, the unit is composed of a steel tube coated with materials which emit infrared waves, electric heating wire, insulating material, and connecting elements. The heating wire is placed in the centre part of the steel tube and is packed around with magnesium oxide (MgO) powder which is a good heat conductor and electrical insulating material.

This kind of radiator can be made in various sizes to suit different oven chamber widths and in two types of configuration, straight and U-shaped. The U-shaped radiator is convenient for installation and maintenance, but its cantilevered tube should be supported in the baking chamber. The straight tube is more stable and can be supported by the two sidewalls of the chamber.

Metal duct radiators are very popular in electric ovens as they have high mechanical strength, a long service life, and can be properly sealed and are easy to replace by removing the outer cover (or door) of the chamber sidewall and pulling out the tube. Care should be taken to prevent the tube coating material peeling off, resulting in food contamination.

The temperature of the medium governs the emission of thermal radiation. A suitable working temperature of the metal tube with infrared-coatings is under 600°C, which ensures a high thermal radiation efficiency.

14.3.1.2   Silicon carbide duct radiators

This radiator is designed with a silicon carbide tube instead of a steel tube, with an electrical heating wire inside. Since silicon carbide is not an electrical conductor, it is

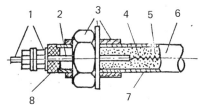

Fig. 14.5   Magnesium oxide–metal duct radiator (1) Wire terminal, (2) Wire connecting rod, (3) Installation device, (4) Heater wire, (5) MgO powder, (6) Coating material, (7) Steel tube, (8) Ceramic insulator.

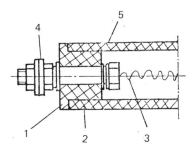

Fig. 14.6   Silicon carbide duct radiator [15] (1) Ceramic insulator, (2) Silicon carbide tube, (3) Heater wire, (4) Terminal, (5) Coating material.

unnecessary to pack the tube with insulating material. Fig. 14.6 illustrates the construction of a silicon carbide duct radiator.

Experiments show that silicon carbide is a good material for emitting infrared electromagnetic waves which match the absorption characteristics of the basic ingredients of bakery foods, such as flour, sugar, vegetable oil, and water, resulting in high heat efficiency. Table 14.7 shows the high emissivity of silicon carbide from different surfaces at 600°C.

Compared with the magnesium oxide-metal duct radiator, the silicon carbide duct radiator is simpler to manufacture, lower in cost, and more durable. It gives higher radiant energy, but it has higher heat inertia (slower to respond to temperature changes) and is relatively fragile mechanically.

14.3.1.3   Silicon carbide rod radiator

Fig. 14.7 shows another type of silicon carbide radiator which is made from high purity silicon carbide powder bound together by an organic bonding agent. The rod is shaped by extrusion, then further treated with silicon carbide and recrystallized at high temperature. This is a type of non-metallic direct fired radiator with no heater wire. The central part of the radiator, the silicon carbide rod, is supported by the side terminals and acts as the heater, and is highly efficient in emitting thermal infrared energy, and quick to respond compared with the duct radiator. This rod considerably saves electrical energy and has excellent baking ability, but it costs more to manufacture and requires skilled installation.

14.3.1.4   Opal quartz duct radiators

This radiator is similar to the silicon carbide duct type (Fig. 14.6) in construction. But the basic tube is replaced by an opal quartz duct with nickel-chromium wire or other

Table 14.7   Emissivity values of silicon carbide [15]

Temperature (°C)	Radiant wavelength (μm)	Emissivity value
600	9.88	0.97
600	1.92–13.90	0.604–0.971
600	2.93–10.00	0.923–0.971

Fig. 14.7   Silicon carbide and radiator.

heating wire inside. The tube is commonly designed with a 18–25 mm diameter and is used as radiator, supporter, and insulator. This radiator has a high and stable emissivity value, $e = 0.92$ as its wavelengths are in the ranges of 3–8 μm and 11–25 μm. With its lower heat inertia, it takes only 2–4 minutes to heat up the oven to thermal equilibrium. Compared with the other radiators, it has a higher heat transfer efficiency from electrical energy to radiation energy (60–65%).

Table 14.8 compares the baking effectiveness of different radiators.

It is known that starch, sugar, egg, vegetable oil, and water have high absorption peaks near the 3 μm wavelength, while at the 4–4.5 μm wavelength there is much lower absorption. Working at 580–600°C, the opal quartz tube emits energy matching the absorption wavelength of the above food ingredients, and this results in high heat efficiency. While the wavelengths emitted by the metal tube radiator are often longer than 4 μm rather than 3 μm, it means that the metal radiators are often coated with infrared-emitting materials. It should be mentioned here that some older metal tubes covered with oxide such as ferric oxide are also highly efficient in emitting infrared waves. That is why old metal radiators are often better than new ones. Table 14.9

Table 14.8   Comparison of baking effectiveness [40]

Radiator	Electric power (kW)	Heating-up time (min.)	Specific energy consumption (kW.h/kg)†
Silicon	11.6	19	0.222
Steel tube	10	25	0.233
Opal quartz duct	7.6	5.2	0.186

†per kilogram of baked products

Table 14.9   Infrared radiation coating compounds with high emissivity [17]

Element	Oxide	Carburet	Nitride	Boride
B	B_2O_3	B_4C	BN	CrB, Cr_3B_4
Cr	Cr_2O_3	Cr_3C_2	CrN	
Si	SiO_2	SiC	SiN	
Ti	TiO_2	TiC	TiN	TiB_2
Zr	ZrO_2	ZrC	ZrN	ZrB_2
Al	Al_2O_3			
Mn	MnO_2			
Fe	Fe_2O_3			
Ni	Ni_2O_3			
Co	Co_2O_3			

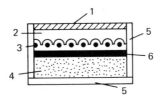

Fig. 14.8   Slab radiator (1) Coating material, (2) Silicon carbide (or metal) plate, (3) Heating wire, (4) Heat insulating material, (5) Heat resistant plate, (6) Electrical insulating plate.

shows some compounds with high emissivity which can be used as infrared radiation coating materials for metal radiators. However, attention should be paid to the adhesiveness of the coating, since any peeling off would cause food contamination.

14.3.2 Slab radiators
As shown in Fig. 14.8, the slab radiator comprises a radiation plate with a heater wire inside, insulating material packed on a heat-resistant plate case, and connecting terminals.

Table 14.10   Radiation data of common radiators [40]

Radiator	Emissivity	Peak wave-length (μm)	Surface temperature (°C)	Heat inertia
Silicon carbide slab	0.84 ~ 0.9	3 ~ 5	250 ~ 300	large
Metal tube	0.85 ~ 0.9	3.5 ~ 6	200 ~ 300	smaller
Ceramic slab	0.9	4 ~ 5.5	235	larger
Ceramic tube	0.9	4.5 ~ 6	230	smaller
Opal quartz tube	0.92	2 ~ 5	150 ~ 350	small

Table 14.11   Experimental comparison results of three radiators [15]

	Baking time and quality		
	Common metal duct	Silicon carbide slab	Silicon carbide duct
Biscuits	5 min. 50 s	2 min. 50 s	3 min. 20 s
Filled bar	10 min. 22 s	5 min. 18 s	6 min
Wire-cut cookie	8 min.		5 min.
Rout press cookie	7 min. 30 s		4 min.
Baking quality	Best for biscuits	Best for cookies	Good for all products
Electric consumption	100%	50%	40%

The radiation plate is commonly made from steel or silicon carbide, covered with infrared radiating materials. Since silicon carbide is an excellent infrared radiation source, further coating with optimum material would result in an even better baking efficiency. At the back of the radiator there is a low conductivity insulating layer allowing heat energy to be concentrated and transferred to the front heating plate, resulting in an increase of electrical energy efficiency.

The power of a radiator unit is 1 to 2 kW for duct radiators and 1 to 5 kW for slab radiators.

The slab radiators are commonly manufactured in square or rectangular units with dimensions suitable for placing across the oven chamber width and for combining two or three units.

Table 14.10 shows radiation data for common radiators. Table 14.11 shows experimental comparison results from three radiators. The ceramic slab and duct radiators (Table 14.10) are not so popular as the others described.

15

Construction and design of tunnel electric ovens

The tunnel oven is often referred to as travelling hearth oven, for the baking process is carried out while the dough piece carrier is travelling. Table 15.1 shows general specifications for some ovens of this type.

A typical tunnel oven comprises a long baking chamber, oven tension end, oven delivery end, band cleaner, and extraction and control systems. Fig. 15.1 illustrates a typical oven.

15.1 OVEN CHAMBER

The baking chamber is the heart of the oven where the dough pieces lose most of their moisture, suffer a large reduction in their density, and change their surface colour owing to the Maillard reaction which produces attractive reddish brown hues.

The electric tunnel chamber consists of a frame supporting the heating elements, rollers, and guiders with heat insulating materials on the top, bottom, and sides. The modern oven chamber is also equipped with heat reflectors inside the chamber, an extraction system, inspection and cleaning doors, and a turbulence and steam injection system.

Most electric tunnel oven chambers are made of mild steel and are continuously welded. For the convenience of shipping and flexibility for various production capacities, the baking chamber is commonly provided in modular units which range from 2 m to 4 m in length. Thus the baking chamber is available in various zone lengths of 2 m, 3 m or 4 m increments to form ovens with a length of from 4 m to more than 100 m. Each zone consists of modules which are joined together on site. Because of the sensitivity of the metal frame to the temperature, its length varies considerably as it heats and cools. To relieve the stress due to expansion and contraction, expansion joints are generally fitted between sections, at the end of each zone.

15.1.1 Oven chamber length

The chamber length is very important, not only for the designer but also for the oven buyer. As the output is predetermined, the length of the oven baking chamber may be

Table 15.1   General specification of some tunnel electric ovens

Model	1	2	3
Dimensions (mm)	$14500 \times 1500 \times 1600$	$17600 \times 920 \times 1000$	$18000 \times 1500 \times 1600$
Installed capacity (kW)	138	150	160
Temperature control range of chamber atmosphere (°C)	20–300	10–300	up to 280
Output (kg/h)	300	500	350–500
Net weight (kg)	6000	6500	7000

calculated according to

$$L_b = \frac{GCT}{60\,RN} \qquad (15.1)$$

where L_b is the length of baking chamber (m), G is the output of the oven (kg/h), R is the number of rows of dough pieces per metre of oven band length, N is the number of rows of dough pieces across the oven band width, C is the number of baked pieces per kilogram, and T is baking time (min.).

Since a typical tunnel band oven consists mainly of three sections, the oven tension end, baking chamber, and oven delivery/drive end, the total oven length L should be the sum of the three section lengths. That is,

$$L = L_t + L_b + L_d, \qquad (15.2)$$

where L_t and L_d are the length of the oven tension and delivery ends respectively.

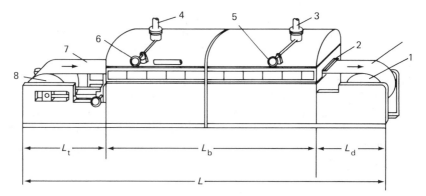

Fig. 15.1   Tunnel electric oven (1) Drive drum, (2) Oven exit, (3), (4) Flues, (5), (6) Damper adjusting handwheels, (7) Oven band, (8) Band tension drum. L_t Length of tension end, L_b Length of baking chamber, L_d Length of delivery/drive end, L total length of oven.

15.1.2   Sectional configuration and dimensions of the baking chamber
As shown in Fig. 15.2, there are two types of baking chamber roof, flat or arched. The flat roof allows a reduced chamber height and lower heat losses, while the arched roof is helpful for steam extraction and prevents the formation of a steam idle corner, but it results in a baking chamber with greater height.

15.1.2.1   Height of the baking chamber
The height of a chamber with flat roof, H, is defined as (see Fig. 15.2)

$$H = L_1 + h + L_2 \tag{15.3}$$

where L_1 is the distance between the top heating element and the chamber roof, L_2 is the distance between the bottom heating element and the chamber bottom, and h is the distance between the top and bottom heating elements.

It is recommended that:

$$L_1 = L_2 = 50 - 70 \text{ mm}$$

The height of a chamber with an arched roof is determined by the equation

$$H = d + L_1 + h + L_2 \tag{15.4}$$

where L_1 is the distance between the top heating element and the bottom of the arched roof, L_2 is the distance between the bottom heating element and the chamber bottom, d is the height of the arched roof, and h is the distance between the top and bottom heating elements. It is recommended that:

$$L_1 = 30{-} 50 \text{ mm}$$
$$L_2 = 50{-} 70 \text{ mm}$$
$$d = 120{-}150 \text{ mm}$$

The magnitude of H is related to the thickness of the products to be baked. The thicker the product is, the larger H will be.

15.1.2.2   Width of the baking chamber
Generally speaking, the width of the baking chamber is dependent on the width of the oven band. Since room is always needed for the installation of the travelling band,

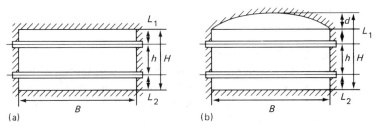

Fig. 15.2   Sectional configurations and dimensions of baking chambers (a) Flat roof, (b) Arched roof.

rollers, guiders, etc., the actual width, B, of the baking chamber (Fig. 15.2) is always larger than the width of the oven band on which the dough pieces are placed. That is

$$B = S + (100 \text{ to } 150) \text{ mm} \tag{15.5}$$

where S is the width of oven band (mm).

15.1.3. Insulation
Heat preservation in the oven is achieved by insulating its chamber walls. Early oven chambers were constructed of brick or stone The brick or stone was both the insulating and oven casing material. Modern oven chamber walls are often constructed from shaped steel frames and insulating layers which are covered and concealed by a thin steel sheet on both sides. Sometimes, a special sheet material composed essentially of heat-resistant fibres and binders is used both as an insulation and constructional component. In this case, an outer covering is not necessary. It is desirable that the covers of the insulating layers are removable for inspection and servicing of the mechanical components located between the inner and outer walls.

Preservation of the oven heat greatly affects the heat efficiency of the oven. If the temperature of the oven outer surface is above 50°C, the quantity of heat loss through the oven walls might be over 20% of the total thermal energy emitted by the electric heating elements [21]. To save energy, the temperature of the outer surface of the chamber walls should be kept as low as possible. However, this could make the insulation layers too thick resulting in huge dimensions of the oven chamber body and an increase in cost for more materials on the chamber frame and cover sheet, etc.

To raise oven heat efficiency and reduce the heat losses through the chamber walls, an oven designer should take the following aspects into account.

(1)   Use high heat resistant, that is, low heat conductivity materials in the insulating layer to reduce heat conduction losses.
(2)   Use materials with high heat reflectivity on the inner walls or reflectors to lower heat absorption losses.
(3)   Reduce the weight and dimensions of the chamber body to reduce the heat emitting area of the outer surfaces of the oven chamber. This limits the thermal storage capacity, giving a fast response and saving in electric power and labour cost by shortening the heating-up time at the start of the baking operation.
(4)   Enhance the oven chamber sealing to avoid unnecessary heat radiation and convection losses.

15.1.3   *Choice of insulation materials*
The properties of the insulation materials are of vital importance for the effectiveness of the preservation of the chamber heat. Generally speaking, materials with a thermal conductivity under 0.2 kcal/m.h.°C are regarded as insulating materials (see Table 15.2).

The oven designer should make the best possible use of the materials of lighter unit weight and lower thermal conductivity for the insulating layer of the chamber.

Table 15.2 Thermal conductivity of commonly used insulating materials [39]

Material	Rock wool	Diatomite	Expanded pearlite	Vermiculite	Glass fibre	Aluminium refractory fibre felt
Density (kg/m^3)	125–300	550	40–120	100–300	100–160	180
Thermal conductivity $(kcal/m.h. °C)$	0.048	0.06	0.024–0.06	0.047	0.035–0.05	0.079

Diatomite, vermiculite, rock wool, expanded pearlite, and aluminium silicate refractory brick are commonly used insulation materials for temperatures under 600°C. Table 15.2 shows their conductivities.

15.1.3.2 Thickness of the insulation layer
The insulation layer of the oven chamber is by no means absolutely heat insulating. Some energy will be lost through the oven walls. The object of a design is to try to reduce this loss, which is also a means of saving energy and money. The thickness of the insulating layer can be calculated according to the principle of heat transfer through a slab consisting of one or several layers under steady state condition. Steady state heat transfer is that when the quantity of heat transferred per unit time is constant; the temperature at any given point in the insulating layer should be constant with time [13]. This condition occurs after the oven has been in operation for some time. For further information on steady state heat transfer, see Refs [13] and [18].

Experiment shows that with the top and bottom of the chamber having the same thickness of insulation, the outer surface temperature of the top is always higher than that of the bottom. Thus, as a general rule, the thickness of the top insulation layer is always larger than that at the bottom of the chamber. It is recommended that the thickness of insulation layers is about 100–150 mm for the bottom, 200–250 mm for the sidewalls, and 250–300 mm for the roof.

15.2 HEATING ELEMENTS

In electric tunnel ovens, there are horizontal bands of electric heating elements above and below the oven band. The heating elements are spaced in zones to meet the requirements of food baking technology.

15.2.1 Installed capacity of heating elements
This is a very important parameter symbolizing the oven output. The thermodynamic equilibrium method is commonly used to calculate how many kilowatts of electric radiators should be installed in an electric oven.

Thermodynamic equilibrium means that the total quantity of heat emitted by the radiators must be equal to the total quantity of heat absorbed and lost by the dough pieces, oven band, and oven chamber walls per time unit. That is,

$$Q = Q_1 + Q_2 + Q_3 + Q_4 \tag{15.6}$$

where Q is the total heat absorbed by objects in the oven chamber (kcal/h), Q_1 is the heat absorbed by the ingredients of the baked dough piece which results in increase of their temperature (kcal/h), Q_2 is the heat absorbed for water evaporation (kcal/h), Q_3 is the heat absorbed by water vapour for oven-heating up to the oven chamber temperature (kcal/h), and Q_4 is the sum of heat losses (kcal/h), which can be calculated by the equation

$$Q_4 = Q_{4a} + Q_{4b} \tag{15.7}$$

where Q_{4a} is the losses through the oven band including all types of oven carriers as they travel outside the oven chamber and emit thermal energy to the lower temperature atmosphere (kcal/h), and Q_{4b} are the heat losses through the walls (kcal/h).

The quantity of the various heats can be calculated by the following equations.

$$Q_1 = \sum G_{1i} C_{1i} \, \Delta t \tag{15.8}$$

where G_{1i} is the weight of all ingredients of dough entering the oven per hour (kg/h), C_{1i} is the specific heat values of the ingredients of the dough (kcal/kg °C), and Δt is the difference in temperature of the dough pieces before and after baking (°C). Table 15.3 lists the specific heat of selected food materials.

$$Q_2 = G_2 q' \tag{15.9}$$

where G_2 is the weight of the evaporated water per hour after entering the oven (kg/h), and q' is the heat of evaporation of water (kcal/h) (Table 15.3).

$$Q_3 = G_2 C_3 \, \Delta t_3 \tag{15.10}$$

where C_3 is the specific heat value of water vapour (kcal/kg °C), and Δt_3 is the difference in temperature of water vapour from 100°C to the temperature of oven chamber (°C).

$$Q_{4a} = G_{4a} C_{4a} \, \Delta t_{4a} \tag{15.11}$$

Table 15.3   Specific heat of selected materials and the heat of water vaporization [39]

Materials	Water	Water vapour	Flour	Sugar	Steel band
Specific heat (kcal/kg °C)	1	0.48	0.5	0.4	0.12
Water vaporization heat (kcal/kg)	539				

where G_{4a} is the total weight of the oven carriers entering the chamber per hour (kg/h); for tray ovens, it is the weight of all trays and driving chains entering the oven chamber per hour, while for band ovens, it is only the weight of the band, either a steel sheet band or a wire mesh band; C_{4a} is the specific heat value of the oven carrier (kcal/kg °C), and it is about 0.12 kcal/kg °C for steel carriers; Δt_{4a} is the temperature difference of the carrier before and after heating up, that is the difference in temperature of the carrier at the entrance and exit of the oven chamber (°C).

As previously stated, Q_{4b} is the heat losses through the oven chamber walls. Since the wall temperature is obviously higher than the temperature of the atmosphere in the oven chamber, there must be natural heat convection. As any object with a temperature higher than 0 K emits energy (radiation), so the oven walls transmit heat to the surrounding space (that is, heat exchange between the oven wall and the surrounding atmosphere) in two ways, natural convection and radiation. That means, the heat losses through the oven walls are made up of two parts; one is lost by natural convection, the other part by radiation. The heat losses of the two parts are related to the temperature of the oven chamber atmosphere, and can be calculated [21] from

$$Q_{4b} = h_c A(t_1 - t_2)^{1.25} + 4.88Ae\left(\left(\frac{T_1}{100}\right)^4 - \left(\frac{T_2}{100}\right)^4\right) \tag{15.12}$$

where A is the total outer surface areas of the oven walls (m^2), h_c is the convection heat-transfer coefficient (it should be about 2.2 kcal/m^{2}h °C in simplified calculation), t_1 is the average temperature measured practically at the outer surfaces of the walls (°C), t_2 is the temperature of the oven chamber atmosphere (°C), T_1 and T_2 are the absolute temperatures of the oven wall outer surfaces and the oven chamber respectively (K), and e is the emissivity of the oven wall outer surfaces.

Since the power is directly proportional to the square of the voltage, the practical installed capacity of the electric heating elements (radiators) should be greater than the calculated result to prevent insufficient baking if any voltage fluctuations occur and to allow an increase of oven output and some changes of conditions. The practical installed capacity or the electric power of the radiators to be installed in the oven chamber, can be calculated from

$$P = \frac{QK_2}{860\,K_1} \tag{15.13-1}$$

where P is the installed capacity of the electric heating elements (kW), K_1 is the correction coefficient of voltage fluctuations (Table 15.4), and K_2 is the power reserve coefficient which can be 1–1.3.

When the necessary data are not available or a simplified calculation is needed, the

Table 15.4   Relationship between voltage (V) and correction coefficient (K_1) [21]

V	240	230	220	210	200	190	180	170
K_1	1.19	1.093	1	0.911	0.826	0.746	0.67	0.6

oven installed capacity can be estimated according to the heat efficiency. That is:

$$\eta_{oven} = \frac{\text{Efficient heat}}{\text{Emitted heat}} = \frac{Q_1 + Q_2 + Q_3}{Q}$$

therefore $Q = \dfrac{Q_1 + Q_2 + Q_3}{\eta_{oven}}$

therefore

$$P = \frac{Q}{860\,\eta_{oven}} = \frac{Q_1 + Q_2 + Q_3}{860\,\eta_{oven}} \qquad\qquad (15.13\text{-}2)$$

where η_{oven} is the oven heat efficiency (generally above 50%). When highly efficient infrared (actually far-infrared) radiators are employed, the η_{oven} can be above 70%, or even up to 80%.

15.2.2   Surface temperature of the radiators

Equation 14.3, $Q/A = \sigma e T^4$, indicates that the energy Q emitted from surface A is directly proportional to the fourth power of the absolute temperature. Thus, the higher the surface temperature of the radiator, the greater the heat flux will be. But as the temperature increases, the wavelength becomes shorter. Therefore, increase of the radiation in the range of infrared wavelength cannot be achieved by increasing the temperature of the radiator surface. To take full advantage of the infrared heating, it is necessary to control the temperature of the radiators to ensure that the wavelengths emitted are mainly in the infrared range and can be strongly absorbed by the food materials. Experiment shows that the surface temperatures of most radiators are controlled in a range between 400 and 600°C, with some below 800°C and others below 400°C for high heat efficiency.

12.2.3   Radiation distance

In the oven chamber, the top heating elements heat up the dough pieces directly by radiation. The radiation distance is the distance between the centre axis of the top radiator, or the heating surface of the top slab radiator, and the surface of the oven band or the baking tray. The magnitude of the distance greatly affects the infrared radiance. In a well-designed oven chamber, with duct or slab radiators, the radiant energy is inversely proportional to the (1.25–1.5)th power of the distance for many reflections [39].

Table 15.5 shows that the radiance decreases with increase in distance. Experimental results indicate that the shorter the radiation distance, the greater the radiance and heat efficiency will be, and the more significant the non-uniform baking quality resulting, especially for duct radiators. But, as the distance decreases to a certain degree, the rate of the radiance increase will obviously slow down. Conversely, the greater the radiation distance, the lower the radiance and baking temperature will be, which also results in increased baking chamber dimensions. But it can lead to well-distributed radiation. In principle, the shorter the radiation distance is, the more compact the structure will be, but always under the preconditions of uniform radiation, good baking quality, and convenient operation.

Table 15.5 Relationship between heat radiance and radiation distance for a
selected radiator [39]

Radiation distance (mm)	180	250	500	750	1000
Heat radiance $(\text{kcal/m}^2\text{h})$	1100	620	230	90	60

Baking uniformity is very important, and much more attention should be paid to
the designing of the stationary tray oven with shelf doors. For the tunnel ovens, as the
dough piece carrier is moving all the time in the oven chamber, there will be no
significant non-uniform baking problem to be considered. The thicker the product,
the larger the distance will be.

The radiation distance can be shortened to 50 mm for biscuit and cookie ovens if
the oven band travels relatively fast. In this case, it results in better heater efficiency.

Fig. 15.3 is a sectional view of a tunnel electric oven. The recommended data of
radiation distances are as follows.

$$\text{Bottom radiation distance } A = 30 - 40 \text{ mm (for slab radiator)}$$

$$A = 60 - 70 \text{ mm (for duct radiators)}$$

$$\text{Top radiation distance } d = B + C \text{ (}B \text{ is about } 2A\text{)}$$

$$d = \text{about } 200 \text{ mm}$$

where C is the height of the product before baking (mm), and B is related to the
configurations and changes of the products during and after baking. Table 15.6 shows
the radiation distances for some selected foods.

It should be mentioned here that the uniformity of baking quality is also related to
the distances between individual radiators.

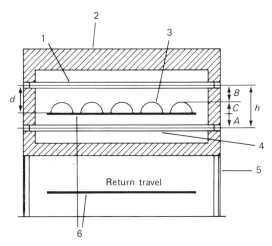

Fig. 15.3   Radiation distance (1) Top radiator, (2) Insulation layer, (3) Product, (4) Bottom
radiator, (5) Oven frame, (6) Oven band.

Table 15.6   Radiation distances suitable for selected foods [39]

Food product	Biscuits	Bread (100g)	Moon cake	Deposited cake
Top radiation distance (mm)	70–120	100–180	100–140	150–180
Bottom radiation distance (mm)	50–70	50–70	50–70	50–70

15.2.4   Radiator arrangement

15.2.4.1   Distance between two radiators

In the oven chamber, the duct radiators are arranged parallel to one another across the top and bottom of the oven band. Fig. 15.4 shows the temperature distribution in an oven chamber when the top radiation distance and the distance between two ducts are 200 mm, Fig. 15.5 indicates that the changes in the temperature distribution curves respond to the variation of the distances between radiators. In bakery ovens, the distance between two ducts is about 150– 250 mm.

For slab radiators with no great difference between length and width dimensions, there will be no great difference in the radiation distribution in the two directions, so that the baking temperature is generally uniform (as shown in Fig. 15.6). It is recommended that the distance between two slab radiators should be about 30– 50 mm.

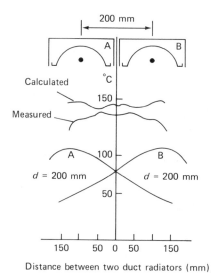

Fig. 15.4   Temperature distribution of two parallel radiators [39].

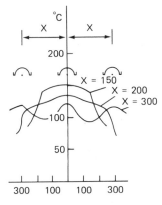

Fig. 15.5   Temperature distribution with slight changes in radiator position near the reflector focus [39].

15.2.4.2   *Arrangement of heating zones*

In electric ovens, the heating elements (radiators) are arranged above and below the oven band to heat the products from the top and bottom. The top and bottom baking temperatures are often referred to as face fire and bottom fire respectively.

For the convenience of radiator replacement, tunnel ovens are commonly equipped with duct radiators rather than slab radiators. There are three radiator arrangements, even installation, grouped installation, and technological installation.

Even installation is the arrangement in which the radiators are installed at equal intervals to provide uniform radiation throughout the chamber. This arrangement is widely used in electric ovens.

In grouped installation, the distances between the groups give pulse-distributed baking temperature. This arrangement is suitable for long tunnel electric ovens. The long oven chamber is heat insulated and has a uniform temperature and very little convection. As the dough pieces heat up some water is transformed into vapour, which tends to form thin films covering the dough pieces and blocking the escape of vaporized moisture from the dough. In this case, forced convection is often needed to prevent the formation of thin vapour films on the dough pieces. The application of the grouped arrangement results in many small pulse-like 'up-and-down' temperature

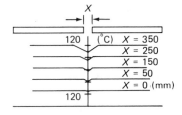

Fig. 15.6   Temperature distribution of slab radiators mounted at different distances apart.

zones. These temperature differentials cause accelerated fluid flow (convection) so that the vapour film is prevented from forming. This gives higher heat efficiency. Experiments have shown that the temperature pulsed distribution can also be achieved when the distance between radiators is larger than 300 mm in even installation.

In 'Technological installation' the radiators are installed according to the baking technology for one particular bakery product, so that the oven is specifically designed for a certain type of product. For instance, the baking process for bread consists of three stages.

In the first stage, after the bread dough pieces enter the oven chamber, the dough should be baked under conditions of lower temperature but higher relative humidity (60–70%). Here, the temperature of the face fire is generally below 120°C. The bottom fire temperature should be high, but not above 250–260°C. In this way, the surface of the bread dough is prevented from forming a hard shell too soon, as this would result in limitation of the specific volume of the bread.

In the second stage, the temperature of the centre of the dough reaches 50–60°C, and the baking temperature can be set at the highest level, with the face fire temperature about 270°C and the bottom fire not higher than 270°C. After this stage, the bread volume is established.

In the third stage, the oven temperature can slowly fall to 180–200°C for the face fire and 140–160°C for the bottom fire, allowing the bread to form a reddish crust and to strengthen the flavouring without being burnt.

For biscuit baking, the electric ovens are commonly designed with three or four heating zones of which the first three are heating zones where the temperature in each zone can be regulated between 200 and 300°C. The last zone is a heat reserve zone in which the heating system does not operate, and the temperature of the atmosphere in the oven is maintained by the biscuits themselves, thus allowing the biscuit temperature to fall gradually. A sharp decrease in temperature could cause cracks or deformations in the biscuits as they travel directly out of the high-temperature heating zone into cold ambient air without passing through a heat reserve zone.

Some modern ovens consist of more zones (up to 6–8) to make the oven temperature control more suitable for more types of bakery product.

15.2.4.3 Power distribution
In tunnel ovens, the top radiators emit radiation energy directly onto the dough. The bottom radiators have to heat the dough bottoms through trays or the oven band. As hot fluids always tend to rise, the installed power for the bottom fire is generally greater than that for the face fire. In tunnel tray ovens used specially for biscuits, the top power is about 75–80% of the bottom power. In tin ovens used specially for bread, the top power is 50–60% of the bottom power. In universal ovens, the power can be equally distributed between the top and bottom and will be regulated according to the different products by means of the oven temperature control system.

15.3.1 Temperature control
It is difficult to achieve an optimum power distribution, especially for those ovens which are employed for several different products. Even for the specific ovens the

temperature still needs to be changed to suit variations in product configuration and size, so that all electric ovens are equipped with an automatic control system including heating controls, heat ratio controls, and alarm monitoring.

Temperature control is the control of the oven chamber atmosphere temperature and the radiator surface temperature. It can be achieved by means of regulating the electric voltage. Since power is directly proportional to the square of the voltage, and the resistance of the heating wire in the radiator is constant, variation of the voltage results in changing the heating power.

Thyristor controlled voltage regulators which allow stepless regulation under load conditions are commonly used in oven temperature control systems. The voltage changes cause an increase or decrease in the surface temperature of the radiators, which directly affects the temperature of the chamber atmosphere. The baking atmosphere temperature is monitored by an automatic control system. When the atmosphere temperature is below or above the preset temperature, this system automatically corrects it, thus maintaining the oven chamber temperature at a substantially stable level.

Another temperature control method changes the number of heating elements used. The radiators are switched in groups. At the beginning, all the radiators are switched on to heat the chamber rapidly. As thermal equilibrium is reached, some radiators are automatically switched off. When the oven chamber is cooled down by loading, a suitable number of radiators will be switched on automatically to make up the necessary power consumption. This method uses the advantages of infrared radiation to the full by keeping the radiators always in the optimum emission state. It gives simple temperature regulation and maintenance, allows saving of energy, provides longer radiator service-life, and lowers oven costs. However, its precision of control is lower than that of the voltage regulation method.

15.4 Reflection devices

Parabolic and flat reflection devices are commonly used in tunnel electric ovens. The reflecting efficiency of a parabola is affected by changes of the position of radiators, but the flat reflector is not so obviously affected.

Fig. 15.7 shows that a parabola reflector reflects parallel rays when the radiator is at the focus of the parabola, and reflects converging or diverging rays when the radiator is deviated slightly from the focus.

It is better to use a parabolic reflector for baking ovens since food dough pieces are often of small size and have a relatively simple surface. Parallel heating rays give

Fig. 15.7   Heat ray distribution with slight changes in radiator position near the reflector focus [39].

reasonably uniform heating. The infrared heating rays absorbed by the dough comprise those direct from the radiator and the parallel rays reflected by the parabolic surface.

As shown in Fig. 15.7, the best reflecting effect is produced when the radiator is placed at the focus of the parabolic radiator. But the radiators used in tunnel electric ovens are not point illuminants but often duct elements with a certain volume across the oven band, so that even if the axis of the duct coincides with the parabola focus its heating surface is not at the focus. Thus, the parabola reflector does not reflect the desired parallel rays. For more parallel rays to be reflected, it is suggested that the axis of the duct be moved out of focus at a distance of 3/8 of the diameter of the duct [21].

Parabolic reflectors are installed only on the roof inside the baking chamber. The sidewalls and oven bottom are equipped with flat reflectors. Even though the flat reflector cannot reflect parallel heating rays, it does reduce heat losses through the chamber walls by absorption and conduction, resulting in an increase of the baking atmosphere temperature and heat efficiency. As mentioned in Chapter 14, reflectors in the baking chamber are commonly made of polished aluminium or stainless steel for which the reflectivity values are above 0.86. After a period of usage, some vapours and oil fumes will pollute the shiny surface of the reflector and result in reduced reflecting efficiency, so that it is necessary to clean the surface periodically to restore high reflectivity.

15.5   Oven flues

In the baking process, a quantity of water vapour escapes from the dough. If the vapour is not rapidly extracted, it may form a thin film on the surface of the dough pieces, hindering the newly vaporized moisture to continue to escape from the dough. Water vapour has absorption zones in the regions 3–7 μm and 14–16 μm of the infrared wavelength, which attenuates heat transmission to the vapour so that there is a decrease in heat efficiency. Because of this, vapour extraction flues are commonly employed in tunnel electric ovens to promote the flow of the water vapour and liquid so that the formation of a vapour film is prevented and the heat efficiency is increased. Any oil smoke is also expelled at the same time.

Generally speaking, tunnel ovens have a high production output which corresponds with a great quantity of water vapour and oil smoke, and this mixture should not be discharged into the oven room but outside the building. Two parameters are invariably necessary, that is, the diameter and height of the flue.

The diameter of the flue can be calculated from [39].

$$d_1 = \left(\frac{B_j n V_y (t_o + 273)}{3600 \times 273 \times 0.785 W_o} \right)^{1/2} \tag{15.14}$$

where d is diameter of the flue (m), $B_j n V_y$ is the quantity of vapour discharged per hour (m^3/h), for bakery ovens,

$$B_j n V_y = \frac{G S_v}{N} \tag{15.15}$$

where G is the weight of the vapour expelled per hour (kg/h), S_v is specific volume of vapour at the discharge temperature, $S_v = 2.3$ m^3/kg for water vapour at 220 °C, as oil smoke can be neglected, and N is number of flues.

For equation (15.14), W_o is velocity of the vapour at the flue outlet (m/s), W_o is about 6–10 m/s at full load and free convection, and 2.5–3 m/s at minimum load [39]; t_o is temperature of the fluid vapour and smoke at the outlet of the flue (°C),

$$t_o = t_i - t_d \tag{15.16}$$

where t_i is the temperature of the vapour at the flue inlet (°C) (it is about the temperature of oven chamber atmosphere), and t_d is the drop in temperature in the flue (°C).

$$t_d = H \, \Delta t_d \tag{15.17}$$

where H is the height of the flue (m), which is dependent on the height of the building above the oven; Δt_d is the drop in temperature of the vapour in the flue per metre (°C/m). For a stainless steel flue with insulating layers, Δt_d should be 3–4°C/m for simplified calculations.

The number of flues for electric tunnel ovens is dependent on the length of the oven chamber. An oven 15 m in length is generally equipped with two flues, each of which is provided with a damper to regulate the quantity of vapour discharged into the inlet of the flue. Theoretically, the flue inlet should be placed in the high temperature zones and delivery end, as at the feed end more vapour is wanted to prevent the dough piece from forming a shell before it is wanted, which would affect the specific volume and cooking quality of the dough centre in the bread baking process. For this reason, some bread ovens are equipped with special ducts inside the oven chamber to introduce steam from the high temperature zones where a great quantity of moisture is vaporized, into the feed end of the oven chamber where more steam is needed to reduce the rate at which the dough surface dehydrates. Some ovens are equipped with a special device to inject steam into the feed end of the chamber for the same purpose.

Modern oven flues are made of stainless steel sheet suitably overlapped. Each flue has an outer casing of high thermal resistant materials (Table 15.2) to prevent a sharp drop in the temperature of the discharged vapour, leading to condensation within the flue which might then return to the chamber and fall down on to the baking dough pieces. It also prevents the high temperature flue from adding more heat to the already hot oven room.

The flues should, if possible, be vertically installed and run up through the roof of the building above the oven. They should be kept separate from each other. Otherwise, they could affect the flue discharge characteristics and result in a feedback of pressure differentials in the baking zones.

Electric tunnel ovens, unlike others, need no fans. The extraction force produced by the height of the flue is generally enough for the vaporized moisture and oil smoke to be discharged. The installation of a fan can cost more for maintenance and power since a fan will bring more cold air into the oven chamber and cause more hot vapour to be expelled through the flues.

15.6   OVEN DRIVE SYSTEM

15.6.1   Oven band speed

The speed of the oven band differs for different products. For instance, the baking time is 2–5 minutes for crackers, 8–10 minutes for cookies, and more than 20 minutes for large bread loaves. As the oven length is fixed, the oven band speed is adjusted by stepless shifting mechanisms, and is calculated from

$$V = \frac{L_b}{t} \qquad (15.18)$$

where V is the oven band speed (m/min), L_b is the length of baking chamber (equation (15.1)) (m), and t is the required baking time (min.).

Oven band speed is a very important parameter for the man in charge of the oven and also for the operators of the preceding equipment which feeds the oven continuously at a rate synchronized with the oven.

15.6.2   Choice of oven drive system

Mechanically, oven drives fall into two groups, band-drum drive and chain-tray drive, both including an emergency wind-out system operated either by hand or by separate battery power.

The band-drum system consists of a feed drum which is always the tension drum, a delivery drum which is invariably the drive drum, an endless band supported by rollers, a band cleaning device, and transmission mechanisms with a variable speed motor.

The tension drum gives a constant tension to the band by moving forward and backward. The movement is achieved by means of a pneumatic system for modern ovens or by a handwheel and worm-gears for others. The tension drum allows band expansion and contraction during baking and cooling.

The oven band includes a sheet steel band, either solid or perforated, or a mesh band consisting of wire mesh, continental mesh, or heavy mesh band. [4].

The solid steel band is made from high quality flexible cold-rolled steel sheet with a thickness of about 1–1.5 mm, which allows repeated heating and cooling without undue deformation. The perforated sheet band is made of the same steel as the solid sheet band but is perforated with closely pitched small holes.

Attention should be paid to the band joints which can be either welded or riveted. In any case, the joints must ensure that the band goes correctly round the drum and runs straight.

Mesh bands are made from woven wire of about 1.5–2.0 mm in diameter. In mesh bands, the wire mesh band is the most loosely woven, whilst the heavy mesh band is the most closely woven one. Subsequently, their weight and service-life is in the order of heavier and longer life from wire mesh to continental mesh to heavy mesh band.

Comparatively, the weight of any type of band, either steel sheet band or mesh band, is much less than the sum weight of the trays and chains, so that the heat absorption and heat losses of the band in baking chamber and during its return travel outside the chamber are significantly less than for the tray-chain system.

Among the different bands, the perforated band is the most expensive, but it demonstrates advantages both over the solid sheet band and the mesh band by virtue of its weight, energy, economy, and its feasibility for many different bakery products. The holes also assist water vapour and fermentation gases to escape, as with the mesh band.

The cheapest band is perhaps the wire mesh band. However, it also has the shortest service-life.

Generally, the solid steel sheet band is the most popular in bakery ovens because of its flexibility in the production of many different products. This is very important for many middle and small scale bakery factories, and also because of its lightness of weight, energy saving, and long service-life, even though it costs more than the mesh band when purchased.

The so-called 'chain-tray drive system' generally consists of two or four endless chains driven by respective chain wheels which are in turn actuated by a variable speed motor (Fig. 15.8) through a reduction box. There are pushing bars or claws on the chains at intervals equal to the tray dimension along the oven length to keep the trays moving forward.

This type of drive is often used for deposited bakery products which cannot be accepted by any type of baking band. The trays may be designed with a certain number of moulds on a thick cast metal slab or made of mild steel plate with a folded structure. In any case, the trays and driving chains are much heavier than the previously discussed band carriers, so that the oven heat efficiency is lower, owing to the high heat absorption in the oven chamber and greater thermal losses as they travel back to the oven feed end. Therefore this type of baking carrier is generally used in specific ovens such as tinned bread ovens and deposited cake ovens.

15.6.3 Band-drum drive design
Two parameters are necessary in band-drum drive design, the power for driving the oven band and the diameter of the drive drum.

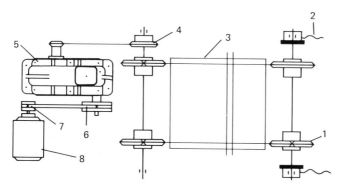

Fig. 15.8   Chain-tray transmission system [15] (1) Chain wheel, (2) Tension drive, (3) Tray, (4) Drive chain wheel, (5) Reduction box, (6), (7), Belt wheels, (8) Motor.

15.6.3.1   *Calculation of the drive power*
The band drive drum is driven by a motor via gears. The necessary power for driving the drum is the essential prerequisite of all calculations of motor power, transmission shaft, and gears.

For drive power, it is necessary to begin with the condition of normal transmission of the oven band.

Fig. 15.9 shows the distribution of tensions and resistances of the band, which is divided into straight and curved sections connecting to each other in turn. Point (1) is the point with the least tension, which is generally the wind-out point of the drive drum. The drive drum should be placed at the oven delivery end; at the feed end is the tension drum which provides the turning point for the oven band. This arrangement ensures that the upper half of the band is in tension so that the dough pieces can be conveyed smoothly through the oven chamber. For normal transmission conditions of the band, we start with the tension and resistance of each point (Fig. 15.9). We start with the band tension S_1 at the wind-out point of the drive drum, along the direction of band, moving step by step until the oven band tension S_n is reached. Finally, the relationship between S_1 and S_n is given by

$$S_n = F(S_1) \tag{15.19}$$

where S_n is the band tension at the wind-in point of the drive drum (kg), and S_1 is the tension at the wind-out point (kg).

According to the principle of mechanical transmission, the condition for band normal transmission is

$$S_n \leqslant S_1 e^{fa} \tag{15.20}$$

where e is the base of natural logarithms, e $= 2.718$, f is the friction factor, and a is angle of contact. As shown in Fig. 15.9, $S_n = S_4$, $a = 180°$. Thus, the tensions at each point of the band can be written as

$$S_2 = S_1 + W_{1-2} \tag{15.21}$$

$$S_3 = S_2 + W_{2-3} + W_g \tag{15.22}$$

$$S_4 = S_3 + W_{3-4} \tag{15.23}$$

where S_1 is the tension at the wind-out point of the drive drum (kg), S_2 is the tension

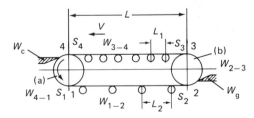

Fig. 15.9   Distributions of tensions and resistance of an oven band [15] (a) Drive drum, (b) Tension drum.

at the wind-in point of the tension drum (kg), S_3 is the tension at the wind-out point of the tension drum (kg), S_4 is the tension at the wind-in point of the drive drum (kg), W_{1-2} is the resistance of the travelling band when free from any load (in return travel outside the oven chamber) (kg), W_{3-4} is the resistance of the travelling band in the loaded section (kg), and W_g is the scraper resistance (kg) which can be measured practically.

Combining equations (15.21), (15.22), and (15.23), gives

$$S_4 = S_1 + W_{1-2} + W_{2-3} + W_g \tag{15.24}$$

The resistance of the band in return travel can be calculated from

$$W_{1-2} = (q_b + q_r)L\omega_r \tag{15.25}$$

where q_b is the weight of band per metre (kg/m), and q_r is the weight of the supporting rolls per metre length in band return travel section (kg/m),

$$q_r = \frac{G}{L_2} \tag{15.26}$$

where G is the weight of each supporting roll (kg), L_2 is the interval of the supporting rolls in the band return section (m), $L_2 = 3$ to 5 m typically, L is the length of the linear part of the band which is equal to the centre distance between the two drums (m), and ω_r is the resistance factor. $\omega_r =$ about 0.025 is recommended for dry and clean working conditions.

The resistance of the tension drum can be calculated from

$$W_{2-3} = K_2 S_2 \tag{15.27}$$

where K_2 is the resistance factor, as the angle of contact of the band is about $180°$, $K = 0.03$ to 0.06.

The resistance of the band in the loading section can be calculated from

$$W_{3-4} = (q + q_b + q_{ro})L\omega \tag{15.28}$$

where q is the weight of the products per metre of oven band, which is dependent on the type of products and the width of the band (kg/m), q_{ro} is the weight of the supporting roll per metre in the band loading section (kg/m).

$$q_{ro} = \frac{G}{L_1} \tag{15.29}$$

where G is the weight of each supporting roll (kg), L_1 is the spacing of the supporting rolls in the band loading section. It is recommended that $L_1 = 1.5$ to 2 m while q_b, L and ω are as in equation (15.25).

By simultaneous equations (15.20), (15.24), and (15.27), tension S_1 and tension S_4 can be calculated.

The necessary peripheral force F (kg) for driving the oven band by the drive drum can be calculated from

$$F = S_4 - S_1 + W_{4-1} + W_c \tag{15.30}$$

Empirically,

$$W_{4-1} = (0.01 \text{ to } 0.02)(S_1 + S_4)$$

Therefore,

$$F = S_4 - S_1 + (0.01 \text{ to } 0.02)(S_1 + S_4) + W_c \qquad (15.31)$$

where W_{4-1} is the resistance of the drive drum (kg), and W_c is the resistance of the scraper (kg), which can be measured practically.

The necessary power for actuating the drive drum can be calculated from

$$N_t = \frac{FV}{102 \, \eta} \qquad (15.32)$$

where N_t is the theoretical necessary power for driving the drum (kW), F is the peripheral force of the drive drum (kg), V is the linear velocity of the oven band (m/sec.), and η is the mechanical efficiency of the transmission system, which is about 0.7–0.8 for ovens.

Considering the practical situation such as the effect of inertial force and possible overloading, the practical drive power N_p is often larger than the theoretical power N_t. That is,

$$N_p = 1.4 \, N_t. \qquad (15.33)$$

15.6.3.2 Drum diameter

To prevent the band from permanent deformation, the diameter of the drum should be calculated under the limit condition so that the outermost surface of the steel band reaches its yield limit. From this point of view, the minimum diameter of the drum (both drive and tension drums) can be calculated from

$$D_{min} = \frac{E\delta}{\sigma_s} \qquad (15.34)$$

where D_{min} is the minimum diameter of the drum (mm), E is module of elasticity (kg/mm^2), δ is the thickness of the band, and σ_s is the yield limit of the steel band (kg/mm^2).

The practical diameter of the drum should be about twice the minimum diameter, that is

$$D = 2D_{min}. \qquad (15.35)$$

Equation (15.34) indicates that the diameter of the drum is inversely proportional to the value of the yield limit of the steel band. That is, the higher the quality of the steel sheet, the smaller the drum diameter. For example, if the oven band is made of alloy constructional steel sheet 15Mn/Ti, 1 mm in thickness, the diameter of both drive and tension drum can be calculated by equation (15.34) and equation (15.35).

Since $E = 2 \times 10^4$ (kg/mm^2)

$\qquad \sigma = 40$ (kg/mm^2)

$\qquad \delta = 1$ mm

$$D_{min} = \frac{2 \times 10^4 \times 1}{40} = 500 \text{ mm}$$

$\qquad D = 2 \times 500 = 1000$ mm.

That is, for a 15 MnTi alloy constructional steel sheet band, a suitable diameter of drum is 1000 mm. A diameter much smaller than this would affect the service-life of the band, while a much larger diameter would require greater oven dimensions and would cost more.

For mesh band ovens, the drum diameter should be about 500–600 mm, since mesh bands are much more flexible than cold-rolled steel sheet bands. But the drum should be able to provide enough friction torque for driving the mesh band without considering any deformation of the band.

The oven band is commonly fabricated in thick mild steel plate with welded flanges. Proper attention should be paid to the installation of drums to ensure that the drum face is revolving concentrically with the axis of the spindle within close limits to assist accurate band tracking. Even though band ovens are commonly equipped with a tracking system, it generally only works in a limited adjustment range.

15.6.4 Drive power of the chain-tray system

The power necessary for driving the chain-tray system is dependent on the normal force and friction imparted to the chains. The power of the electric driving motor can be calculated from [39]

$$n = \frac{K_1 N_1}{\eta_1} \qquad\qquad (15.36)$$

where N is the power of the drive motor (kW), K_1 is a safely factor (recommended $K_1 = 1.5$), η_1 is the mechanical efficiency ($\eta_1 = 0.8$ for the system shown in Fig. 15.8), and N_1 is the power of the input shaft of the reduction box. It can be calculated from

$$N_1 = \frac{M n_1}{875 i \eta} \qquad\qquad (15.37)$$

where n_1 is the speed of the input shaft of the reduction box (rev/min), i is the ratio of the reduction box, η is mechanical efficiency ($\eta = 0.95$), and M is the torque of the output shaft of the reduction box (kg/m). It can be calculated from

$$M = \frac{T D_o}{D_d} \qquad\qquad (15.38)$$

where D_o is the diameter of the chain wheel on the output shaft of the reduction box (mm), D_d is the diameter of the drive chain wheel on the driving shaft of the chain-tray system (mm), and T is the torque of the driving shaft of the chain-tray system (kg.m).

$$T = KBf \frac{D}{2} \tag{15.39}$$

where K is the factor for working conditions (recommended $K = $ about 2–3 for high temperature conditions), B is the total weight of dough pieces and the carriers including trays and chains in both loading and return travel sections, f is factor of slide friction between steel parts ($f = 0.05$), and D is the diameter of the chain wheel driven by the driving shaft of the chain-tray system (m).

Part VIII

Candy-making machines

16

Cooking machines

16.1 INTRODUCTION

More than forty years ago, most candy† was hand-made by a manufacturer skilled in the art of candy-making. Today, candy production has been developed from an art to a technology, from labour-consuming work to highly automated mechanization, and from a small scale workshop to large-scale continuous and highly efficient production lines except for some special varieties.

Justin J. Alikonis pointed out [7] that there are over 2000 kinds of candies which have hundreds of different variations. The difference in candies is essentially due to the different formulas and the different ways of removing a varying amount of moisture and controlling or preventing the formation of crystals as they are cooked. For example, hard candies owe their hardness to the removal of all but about 1 % of their moisture by heat and vacuum cooking. For the gamut of chewy candies (caramels, toffees, and nougats) and soft candies (creams, fudges, marshmallows, and jellies), the moisture content is higher. [7]

Candy production comprises five stages: cooking, cooling and tempering, shaping, cooling, and packing.

The cooking process is generally a three-phase working cycle, that is, weighing, dissolving, and cooking, of which each phase is controlled separately.

16.2 COOLMIX SYSTEM

The Coolmix is an automatic weighing and mixing installation comprising a stainless weighing vessel, a stainless lower reservoir, stainless pipes, and a control system with a facia panel (Fig. 16.1, 16.2).

All raw materials are weighed and thoroughly mixed in the upper weighing vessel (1) and then discharged into the lower reservoir (2) in which there is a high-speed

---

†'Candy' is crystallized sugar, in both the USA and the UK. But USA 'candies' are usually called 'sweets' in the UK, though a few such products are known as '——candies'. The word 'sweet' also, in the UK, means the course that follows the main course of a meal, and comprises, say, apple tart, cabinet pudding, or whatever. Colloquially, this 'sweet' course is also known as 'afters'—'What have we got for afters, mum?'

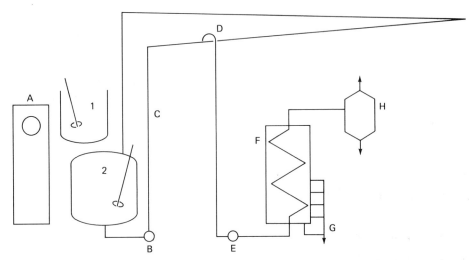

Fig. 16.1   Coolmix feeding one of more dissolving/cooling units via a ring main (Courtesy of ter Braak BV, Netherlands) (A) Coolmix—automatic weighing and mixing installation. Mechanical balance, electrical control. (1) Weighing vessel, (2) Lower reservoir, (B) Circulation pump, (C) Ring main, (D) Take-off to production unit, (E) Feed pump to dissolving/cooker unit, (F) Continuous dissolver/cooker, (G) Condensate, (H) Flash-off chamber.

turbulent agitator to ensure a uniform suspension. The mixture is composed of small sugar crystals suspended in a sugar solution, with or without glucose syrup, which will be fed to one or more dissolving/cooking units either through a ring main (Fig. 16.1) or to an individual production line through a direct feed pump (Fig. 16.2). The discharge of the batch in the weighing vessel is controlled by the level probes in the lower reservoir.

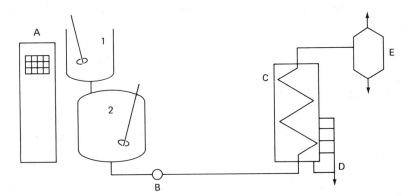

Fig. 16.2   Coolmix feeding an individual production line through a direct feed pump (Courtesy of ter Braak, BV, Netherlands) (A) Coolmix—fully electronic weighing and mixing installation. Electronically controlled balance. Vessel mounted on load cells. Possible connection to computer system (1) Weighing vessel, (2) Lower reservoir, (B) Feed pump with drive, (C) Continuous dissolver/cooker, (D) Condensate, (E) Flash-off chamber.

There are two types of weighing and control system for the Coolmix. One is automatic weighing and mixing with a mechanical balance and electric control, and the other is fully electronic weighing and mixing with an electronically controlled balance. The accuracy over a range from 5–500 kg is ± 0.15 % for electronic execution with a microprocessor. The system allows the operator to maintain control over the entire process cycle and to make adjustments from the control panel if necessary.

The specifications of the Coolmix system is as follows (Courtesy of ter Braak BV, Netherlands).

Capacity of weighing vessel	195 litres
Capacity of lower reservoir	395 litres
Valve for water	1 in
Valve for glucose	2 in
Sugar feeder inlet	200 mm
Electrical consumption	7.5 kVA
Consumption of air	500 Nl/h
Dimensions	1700 × 1250 × 2770 mm
Weight	1200 kg

For different kinds of production line, there are many variations of the Coolmix system such as the Hotmix, Caramix, or Warmmix in single, twin, or triple execution.

16.3 DISSOLVING AND PRECOOKING MACHINE

Weighing, followed by the dissolving process, provides a uniform mixture of measured water, partly dissolved sugar, glucose syrup, and other ingredients, often referred to as 'slurry'. The next procedure is precooking.

Some machines permit the three procedures (dosing, dissolving, and precooking) to be automatically accomplished by a combined unit, such as the Solvomat shown in Fig. 16.3.

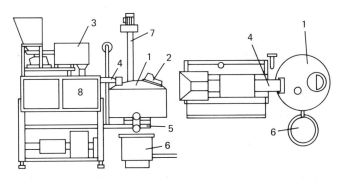

Fig. 16.3 Solvomat unit (Courtesy of BOSCH—Robert Bosch GmbH, Germany) (1) Dissolving and precooking vessel, (2) Inspection window, (3) Crystal sugar weigher, (4) Sugar vibratory tray, (5) Steam inlet, (6) Intermediate container, (7) Vapour discharge, (8) Control cabinet.

16.3.1   Construction

The Solvomat consists of a crystal sugar weigher and vibratory tray, pump systems for water, scrap, and glucose, dissolving and precooking vessel, steam supply, and heater coil devices, all of which are mounted on a sturdy steel frame. The control panel is fixed on the side frame of the machine, providing indicator lamps for weigher, coarse feed, fine feed, and faults on the weigher, off/on buttons for the weigher and the pumps potentiometers for the sugar feed tray, and fine feed, temperature indicator, pump stroke counter (impulse for weigher), and a counter for the number of weighings.

The dissolving vessel is mounted on two supports on the basic frame and consists of inner, outer, and middle chambers, fitted with a lid in which there is a feeding orifice for the crystal sugar, an inspection window, and a hand grip. The inner chamber is equipped with a glucose preheater, and the outer chamber has a heating coil.

The unit is also equipped with an adjustable gear motor and pumps. The rotary slide valve pump is for the glucose syrup, and the valve pump for the water used to dissolve the sugar.

For hygiene, all parts in contact with food materials are made of stainless steel or other food quality materials.

16.3.2   Operation

As shown in Fig. 16.4, the ingredients are individually dosed in a pre-set mixing ratio and are fed into the dissolving vessel where the outer chamber is fed with crystal sugar and water while the inner chamber is fed with glucose syrup. The crystal sugar is

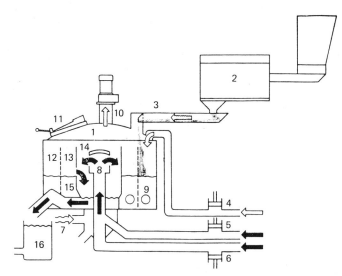

Fig. 16.4   Operation of Solvomat (Courtesy of BOSCH—Robert Bosch GmbH, Germany) (1) Dissolving and precooking vessel, (2) Crystal sugar weigher, (3) Sugar vibratory tray, (4) Water pump, (5) Scrap pump, (6) Glucose pump, (7) Steam inlet, (8) Glucose preheat, (9) Heater coil, (10) Vapour discharge, (11) Inspection window, (12) Outer chambers, (13) Middle chambers, (14) Inner chamber, (15) Sugar solution, (16) Intermediate chamber.

discharged from the sugar silo, weighed in batch, then fed by as vibrator tray, the water being fed by a valve pump into the outer chamber of the vessel. The mixture of sugar and water is heated by the steam heating coil, and the sugar crystals are completely dissolved. The sugar solution is then precooked in the middle chamber of the vessel to a static dry substance (DS) of 80 %. To prevent the vapour from escaping at the sugar inlet, a ventilator is installed on the vapour discharge outlet. In the inner chamber of the dissolving vessel, the glucose syrup is preheated, and then added to the precooked sugar solution. The finished solution is fed continuously into an intermediate container. Other ingredients such as fat, milk, and honey may be added at this stage.

The required dosing accuracy of the individual components is $\pm 0.1\%$. For crystal sugar, this is achieved by gravimetric weight analysis (independent of bulk density). Accurate volumetric dosing is used for all liquid components, which is related to the speed of the pump shaft. Adjustment of the volume and the relationship between components such as water, sugar, and glucose is made by variation in the pump shaft speed.

Table 16.1 gives the specifications of this kind of dissolving and pre-cooking machine (Solvomat).

16.4 PRESSURE DISSOLVER

The pressure dissolver is designed according to the analysis of the relationship between pressure, temperature, and the solubility of sugar. As shown in Fig. 16.5, a higher pressure (higher temperature) results in a higher sugar concentration, that is, higher DS (dry substance), which means lower water consumption to dissolve the granulated sugar and a considerable energy saving in the cooking stage. Fig. 16.6 illustrates the pressure dissolver, and Table 16.2 gives the specifications.

Table 16.1   Specifications of dissolving and precooking machine (Solvomat 126F, Courtesy of BOSCH, Robert Bosch GmbH, Germany)

Output	300–1250 kg/h (adjustable)
Sugar (plus glucose)	Adjustable sugar/glucose ratio 100:30 up to 100:80, with other pumps also up to 100:200
Drive	1 kW (total)
Steam consumption	app. 23 kg steam for 100 of dissolved glucose solution
Steam pressure	Acceptable overpressure 10 bar, working pressure 8 bar (max)
Steam connection	$1 \times R = 1\frac{1}{2}$ in.
Condensate connection	$R = \frac{3}{4}$ in.
Machine dimensions	$2000 \times 1500 \times 2000$ mm
Weight (net)	app. 1500 kg

Table 16.2   Specifications of a pressure dissolver (Courtesy of BOSCH, Robert Bosch GmbH, Germany)

Throughput	sugar/glucose solution 85% DS, app. 1500 kg/h
Working excess pressure	10 bar
Steam room capacity	250 litre
Steam consumption	250 kg/h (with max. performance and one dissolving coil)
Machine dimensions	app. 800 × 100 × 1700 mm (standard version)
Weight	app. 950 kg (standard version)

16.4.1   Construction

The pressure dissolver generally comprises a steam chamber with one or two pipe coils incorporated which are equipped with special vortex baffles to ensure complete dissolving of the sugar crystals inside after a short period. The dissolving temperature is set by the steam control on the cabinet, while the pressure in the dissolving coil is set by means of the pressure retaining valve with its pressure indicator. The product

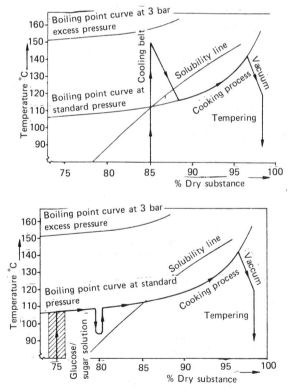

Fig. 16.5   Temperature concentration. Comparison of pressure-dissolving and open-dissolving (Courtesy of BOSCH—Robert Bosch GmbH, Germany).

mixture is pumped by a feed pump which is stepless adjustable to suit different outputs and conditions for a wide range of products.

16.4.2   Operation
As shown in Fig. 16.6, the product mixture, which consists not only of sugar, glucose syrup, and water, but also milk, gelatine, agar, and other ingredients if required, is transferred from the supply container to the cooking coil of the pressure dissolver by an infinitely adjustable gear pump. Within the coil the mixture is heated (without evaporation) to the necessary temperature, which ranges from 108°C (for gelatine products) to 145°C for deposited fondants and high boilings. Different temperatures correspond to different pressures. The pressure, ranging from 0.4 to 4 bar, in the dissolving coil can be controlled by a pressure retaining valve.

The dissolving process has two stages. Firstly, the three-way valve is set to circulation position until the required dissolving temperature has been obtained, that is, the product mixture is returned into the supply container without any loss. Secondly, the circulation flow of product is shut off by the three-way valve, and the dissolved solution is transferred into a supply container or a vacuum cooker for cooling or post-evaporation.

The specific parameters (pressure, temperature, and flow speed) of the process should be in accordance with the product variety and the machine output to obtain a crystal-free solution without any browning of the mass. The pressure dissolving

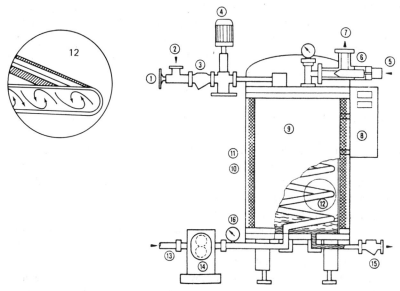

Fig. 16.6   Pressure dissolver (Courtesy of BOSCH—Robert Bosch GmbH, Germany) (1) Steam check valve, (2) Steam supply, (3) Filter, (4) Pressure reducer, (5) Compressed air, (6) Pressure retaining valve, (7) Product discharge, (8) Control cabinet including digital display, mass temperature display, and pump speed indicator, (9) Steam area, (10) Insulation, (11) Housing guards, (12) Coil with vortex baffles (product in vortex motion), (13) Infeed of mixture, (14) Feed pump, (15) Condensate, (16) Pressure indicator.

process is characterized by the reduced water content (high DS) which shortens the time needed for cooking and gives considerable energy saving and economic benefit.

16.5 COOKING MACHINES

Cooking is the basic operation in the candy making process. The objects of the cooking process are to reduce the moisture content of the product mixture, to melt, to solubilize, to caramelize, if necessary, and to invert.

There are many types of cooking machine which may be classified in several ways for the convenience of discussion. According to the working mode, they are batch cookers, semicontinuous cookers, and continuous cookers. According to the method of heating, they come as direct-fired cookers or steam-jacketed cookers. According to the working pressure, they are atmospheric cookers, pressure cookers (steam-injection cookers), and vacuum cookers.

16.5.1 Batch cookers
Two types of batch cooker are widely used in candy production: the vacuum batch cooker and the atmospheric batch cooker.

16.5.1.1 Vacuum batch cookers
The vacuum cookers, either batch or continuous, are designed according to the analysis of the relationships between boiling point and DS of the sugar solution as well as between boiling point and cooking pressure. As shown in Fig. 16.5, the boiling point rises with the increase of DS of sugar solution, but it falls under vacuum conditions. Because of this, the vacuum evaporation process is widely used for the production of hard boilings, which allows the sugar solution to be evaporated at a much lower temperature and in a shorter time without variation in quality but with a clearer appearance and excellent colour and product taste than from atmospheric evaporation.

The vacuum batch cooker is primarily designed for the cooking of hard candy sugar mass. It can also be used for the production of jelly and gum candies as well as some low boiled sweets. Fig. 16.7 shows a typical vacuum batch cooker which is a development of the well-proven Hansella batch cooker. It consists of a cooking chamber, a draw-off kettle attached by suction to the vacuum chamber which is located under the cooking chamber, a special stirrer and its drive system, a vacuum pump with spray condenser, and a control system on the facia.

The cooking vacuum assembly is bolted to a console, by beam construction, on the sturdy stand. The inner part of the cooking chamber is made of copper and the outer of steel. The two parts are joined to form the steam-heating jacket. Inside the cooking chamber, the stirrer is driven variably by a selectable pole motor via a V-belt transmission system, and a sturdy temperature sensor is located at a low position for 'picking up' the temperature of the cooking mass. The conical valve through which the stirrer shaft passes controls the outlet of the cooking chamber to the vacuum chamber/draw-off kettle. The draw-off kettle, made of copper, is equipped with a hinge and an elevating device for convenient discharge of the finished mass.

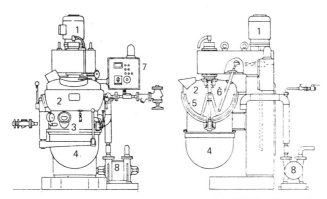

Fig. 16.7 Typical vacuum batch cooker (Courtesy of BOSCH—Robert Bosch GmbH, Germany) (1) Drive motor, (2) Cooking chamber, (3) Vacuum chamber, (4) Draw-off kettle, (5) Stirrer wing, (6) Temperature sensor, (7) Facia, (8) Vacuum pump.

The cooking chamber is fed with the ingredients either manually or by the dosing system of dissolvers described earlier. The special stirrer wings cover the entire heating range and cause an intensive heat transfer to the mass at the heating wall (Fig. 16.8). As the sugar mass reaches the selected temperature, measured by the temperature sensor, the steam supply is switched off automatically. At the same time, a panoramic lamp on the facia panel indicates that the cooking process has been completed, and

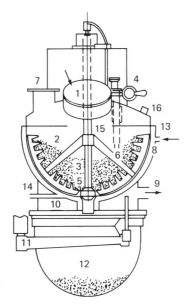

Fig. 16.8 Functional diagram of vacuum batch cooker (Courtesy of BOSCH—Robert Bosch GmbH, Germany) (1) Infeed orifice, (2) Cooking chamber, (3) Valve rod, (4) Valve wheel, (5) Valve, (6) Stirrer wing, (7) Vapour exhaust, (8) Steam chamber, (9) Vacuum connection, (10) Vacuum chamber, (11) Hinge device, (12) Draw-off kettle, (13) Steam connection, (14) Condensate connection, (15) Temperature sensor, (16) Rinse connection.

the pressure in the steam chamber is reduced by a steam pressure relief valve. Subsequently, the vacuum pump is switched on automatically so that the hinged draw-off kettle can be attached to the vacuum chamber by suction. At the same time the valve is raised for the cooked mass to be further evaporated under vacuum conditions. The vacuum evaporation time, that is, the suction time for the kettle, can be selected by setting the height of the valve cone by means of the handwheel (with a scale graduation). The vapours are condensed in the spray condenser of the vacuum pump. The vacuum pump is switched off automatically when the vacuum evaporation has been completed. The draw-off kettle is lowered, hinged, and further tilted for the evaporated sugar mass to be discharged. Table 16.3 gives the specifications of two vacuum batch cookers.

16.5.1.2   *Atmospheric batch cookers*
The earliest atmospheric batch cooker was direct-fired with manual dosing and stirring. The quality of product depended on the skill of the operator.

Table 16.3   Specifications of two vacuum batch cookers (Courtesy of BOSCH, Robert Bosch)

Output (in accordance with the recipe, kg/8h)	app. 1000–1200	app. 1600–2000
Batch size (kg)	up to 40	up to 70
Steam consumption (kg/h)	app. 75	app. 130
As the steam is used in charges the steam generator has to produce (kg./h)	150	260
Stirrer drive motor	1.5 kW gear motor	3.6/4.3 kW pole-selectable
Stirrer wing revolution (rev/min)	80	50/100
Acceptable operation pressure	10 bar	10 bar
Steam chamber contents (litre)	12.7	28
Supply content (litre)	80	150
Vacuum pump drive (kW)	4 kW at 1500 rev/min	4 kW at 1500 rev/min
Water consumption (m^3/h) max. app. (with an intake temperature 15°C)	0.7	0.9
Space requirement including vacuum pump	1800 × 200 × 2100 (mm)	2070 × 2160 × 2400 (mm)
Weight (kg)	950	1250

The modern atmospheric batch cooker, with a wide application of modern techniques, permits the cooking operation to be carried out fully automatically to yield
high-quality products. This group of cookers is specifically designed for the production of soft candies such as caramel, toffee, and jellies under open air conditions. This
type of cooker is similar in construction to the vacuum batch cooker group (shown in
Figs 16.7 and 16.8) but without the vacuum system. The lower part of the cooker is a
steel jacketed kettle (pot, pan) suitable for heating by steam to a maximum pressure of
10 bar. The top part of the cooker is a stainless steel cover which is commonly
equipped with a latched lid, with a transparent inspection window, for manual feeding
of the raw materials. Automatic feed may be accomplished either via a ring main or
directly by the previously described Coolmix, Caramix, Solvomat, or pressure
dissolver.

The cooker is also equipped with a two-speed bronze anchor-shaped stirrer fitted
with special Teflon scrapers to keep the inner wall clean of product. The stirrer is
driven by an electric motor via an enclosed worm gear reduction box. The stirrer shaft
is grooved over the entire length to allow the spindle of the discharge valve to pass
through. The discharge valve may be controlled pneumatically by an activating piston
which lifts the spindle of the valve. The flow rate responds to the discharge valve gap
which is regulated by a handwheel mounted on the top of the gear box.

Technical specifications of a typical atmospheric batch cooker are as follows:
(Courtesy of ter Braak BV, Netherlands).

Working capacity	300 kg/h
Batch capacity	80-130 kg/batch
Steam working pressure	10 bar max
Steam consumption	100 kg/h
Power consumption	7.5 kVA
Stirrer power	4–5.5 kW
Stirrer speed	48–96 rev/min
Dimensions	1140 × 1450 × 2350 mm
Net weight	700 kg

16.5.2 Vacuum continuous cookers

The continuous cooker with vacuum evaporation and batch discharge system
(Fig. 16.9), often referred to as a 'vacuum continuous cooker', is the most popular
equipment used for the production of hard boilings. The continuous cooker is
preceded by a dissolving and precooking machine or pressure dissolver, and is
followed by a cooling and tempering band or a cooling slab or a kneader.

16.5.2.1 Construction

A typical vacuum continuous cooker comprises a continuous cooker, a vacuum
evaporator, and a pair of batch discharge vats (pans, pots, or bowls).

As shown in Figs 16.9 and 16.10, the continuous cooker is designed with a
construction similar to the preceding pressure dissolver but it does not have a three-
way valve or the circulation operation used in the dissolving process. The coil within

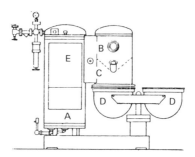

Fig. 16.9 Vacuum continuous cooker (Courtesy of BOSCH—Robert Bosch GmbH, Germany) (A) Steam dome cooking coil, (B) Vapour separation chamber, (C) Vacuum chamber, (D) Vacuum and discharge vat, (E) Control cabinet/facia.

the steam dome is made of copper or stainless steel, and the dome is heated by saturated steam at a maximum pressure 10 bar. The operating pressure depends on the output of the machine and is up to 8 bar with a steam consumption of 200 kg/h, and the output is 750 kg/h or 6 tons/8 h.

The continuous cooker is followed by the vapour separation/vacuum chamber which is of chrome nickel-steel welded to form a combined cylindrical vessel with a jacketted conical bottom and two separate spherical bottom pots. The passage from the vapour separation chamber to the vacuum evaporation chamber is controlled by a two-way valve adjustable for the variation of the flow rate of the cooked product.

Like other parts, such as the chambers described above in contact with food products, the two pots receiving and discharging alternatively are made of either stainless steel or performance-proven copper for food hygiene and a clean attractive appearance. Under the two pots is a turntable used to make possible the use of the two pots alternately in consecutive production.

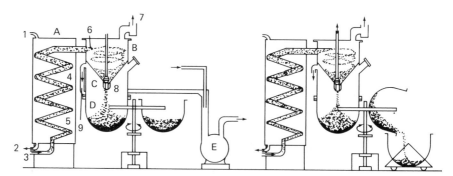

Fig. 16.10 Functional diagram of a typical continuous vacuum cooker (Courtesy of BOSCH—Robert Bosch GmbH, Germany) (A) Steam dome, (1) Steam inlet, (2) Condensate, (3) Sugar solution, (4) Cooking coil (Ni/Cr), (5) Swirling of solution, (B) Vapour separation area, (6) Tangential inlet of cooked sugar solution into vapour, (7) Vapour draw-off (8) Two-way valve, (C) Vacuum chamber, (9) Line to vacuum pump, (D) Vacuumizing and batch discharge vat, (E) Vacuum pump.

The installation is also equipped with an attached control cabinet provided with a facia for automatic or manual control of the steam pressure valve position, temperature of cooking and evaporation, outlet valve setting, level vapour separation, level of vacuum chamber, etc. In addition, this system may be connected to a central monitor to display a flow chart of the continuous cooking plant with a cooling and tempering band.

Table 16.4 gives the technical specifications of some vacuum continuous cookers.

16.5.2.2 *Operation*

The dissolved solution is fed continuously into the cooking coil (4) in the steam dome A (Fig. 16.10) by the feed pump. The sugar solution is 80% DS from an atmospheric dissolver such as Solvomat, or 85% DS from a pressure dissolver. The bottom end of the cooking coil is connected to the pump, and its upper end is linked with the vapour separation chamber. The maximum pressure of steam is up to 10 bar and the operating pressure is up to 8 bar. Since copper and steel are highly efficient metals for heat transfer, the temperature of the sugar mass increases sharply in a very short time. Additionally, vortex elements within the coil are incorporated in the first stage of cooking to achieve optimal heat exchange so that moisture in the solution is vaporized, as in Fig. 16.6(12).

In the middle part of the coil, with continuous heating, extensive evaporation takes place so that the upper end of the coil is full of vapour. When the vapour passes into the vapour separation chamber, it will be exhausted through the extraction outlet of the chamber. By then, the steam cooking phase is finished, and the sugar solution is concentrated to the preset DS, for instance 97%.

At the outlet of the coil, a temperature probe provides signals for the steam pressure

Table 16.4 Specifications of two vacuum continuous cookers (Courtesy of BOSCH, Robert Bosch GmbH, Germany)

Throughput (kg/h)	750	1250 initial DS 80%
Drive motor for sugar pump	0.37 kW 1500 rev/min	0.75 kW 50 Hz
Drive motor for vacuum pump	5.5 kW 1500 rev/min 50 Hz	11 kW 50 Hz
Steam pressure	admissible 10 bar	10 bar
	operation up to 8 bar	8 bar
Steam consumption	200 kg/h max.	app. 380 kg/h
Water consumption vacuum pump at water	750 litre/h	app. 1.25 m^3/h
temperature 15°C	750 litre/h	app. 1.25 m^3/h
Machine dimensions (including vacuum pump) (mm)	2250 × 2410 × 2390	3230 × 1870 × 2500
Net weight (kg)	2045	

regulation system and the product temperature, which ensures accuracy in controlling the sugar mass.

After leaving the cooking coil, the mixture of the boiled sugar mass and vapour are fed via the tangential inlet into the vapour separation section (B) where the boiled mass acquires a good boiling state with expansion of the vapour and further separation from the mass, hence the terms vapour separation or vapour expansion chamber. The sugar mass is collected at the conical end of the vapour separation chamber (B) and the vapour drawn off through the outlet (7). When the preset batch quantity of sugar is obtained, the conical valve (two-way valve) (8) is lifted automatically or manually to allow the sugar mass to flow down into the vacuum chamber where a vacuum of up to 740 mm Hg is generated by means of a vacuum pump. Under the vacuum condition, the sugar mass boils at a lower temperature, the moisture is further evaporated so that the dry substance content of the sugar mass (DS) is further increased, and the mass cooled off at the same time with the heat loss in evaporation. With the increase in DS and the decrease in temperature, the mixture of sugar mass becomes very viscous and is collected in the vacuum/discharge pot. After the preset batch size has been accumulated, the two-way valve shuts off the vacuum chamber. Simultaneously, the vacuum chamber is aerated by the opening of the ventilation valve. As soon as the vacuum is released the filled pot (D) separates from the vacuum chamber and is lowered by its own weight and automatically swivelled 180° into the discharge position by a pivoted actuator. Conversely, the empty pot rotates by 180° exactly into the vacuum position and is sucked up beneath the vacuum chamber in a subsequent fast vacuum generation. Once the preset vacuum is obtained in the vacuum chamber (C), the two-way valve (8) opens automatically and the next batch of the mixture of sugar begins to be concentrated under the vacuum condition while the pot in the discharge position is being emptied onto a cooling slab or into a transporting pan or a cooler. The discharged product is a high viscous sugar mass with residual moisture of about 1%.

16.5.3   Continuous film-cooker
This type of cooker is designed on the principle of a thin-film evaporator for the production of hard milk caramel, soft caramel (toffee, chewable candy), and other products.

16.5.3.1   *Construction*
The standard film-cooker consists of two sections. The first is designed with a cylinder and a vapour separation chamber, and the second has three functional units, a mixing container, a caramelizing container, and a vacuum container (Figs 16.11, 16.12).

The horizontal cylinder in the first section (Fig. 16.12) is designed with an insulated double-jacketed heating unit in which there is a rotor driven by a motor via a clutch. There are two types of rotor. One is shown in Fig. 16.13, which is a closed pipe, made of chrome nickel steel, with milled pockets housing movable Teflon scrapers. Another is shown in Fig. 16.14, which is equipped with Teflon band-scrapers fixed on the centre rotor in steep helical arrangement. A thin film of sugar solution is formed and cooked in the narrow gap between the inner wall of the steam-heated cylinder and the

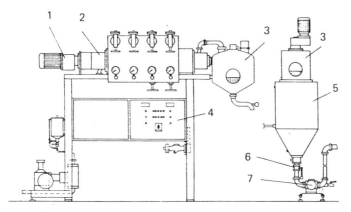

Fig. 16.11   Typical continuous film-cooker (Courtesy of BOSCH—Robert Bosch GmbH, Germany) (1) Rotor drive, (2) Cylinder film-cooker, (3) Vapour separation chamber, (4) Control cabinet/facia, (5) Container, (6) Check valve, (7) Feedout pump.

scrapers of the revolving rotor. Because of this, the installation is often referred to as as a film-cooker or rotor cooker.

The film-cooker is followed by a vapour separation chamber which is similar to that in the vacuum continuous cooker (shown in Fig. 16.10) in construction.

One of the three function units of the second section is the mixing container (Fig. 16.12, A) primarily used for masses where colouring, flavours, or acids are added. It is a steam-jacketed cylinder with a conical bottom connecting with a feedout pump via a check valve. Inside the container, there is as rotary agitator driven by a motor mounted on top of the vapour separation chamber. The passage between the mixing container and upper chamber is controlled by a two-way valve which regulates the flow rate of the product mass.

The caramelizing container is used for products containing milk. It is also a cylinder with a conical bottom, heated by steam coil within the jacket. The agitator with the scraper is used not only for continuously scraping the caramelized product off the inner jacket of the cylinder but also for mixing it with the bulk mass to obtain a uniform caramelized product. The degree of caramelization is controlled by the respective level in the container, that is, the dwelling time.

The vacuumizing container is used for cooling off the product mass and reducing its water content at the same time. The configuration of this container is similar to the preceding two, but without a heating element or agitator. It is connected with the vacuum system as in Fig. 16.10. However, its bottom is not a discharge valve but a valve through which the greatly concentrated product is transferred by means of the feedout pump to a mixing container where the other ingredients are added.

As shown in Fig. 16.11, the continuous film-cooker has also a control cabinet with facia, pumps and varispeed motors for stepless adjustment for a wide range of recipes and throughputs, and a robust frame which supports the basic sub-units including cylinder cooker, vapour separation chamber, and the respective containers.

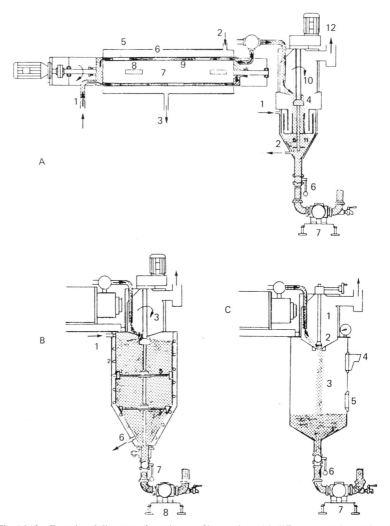

Fig. 16.12 Functional diagram of continuous film-cooker with different containers (Courtesy of BOSCH—Robert Bosch GmbH, Germany) Function: Rotor (1) Sugar solution, (2) Steam supply, (3) Condensate, (4) Feed pump (not shown), (5) Outer jacket/inner jacket, (6) Steam chamber, (7) Rotor, (8) PTFE scraper, (9) Product film, (10) Vapour separation chamber, (11) Mixing container, (12) Exhaust vapours, (A) Function: Mixing container, (1) Steam feed, (2) Condensate, [(10)] Vapour separation chamber, (4) Adjustable flow valve, (5) Agitator blade, (6) Check valve, (7) Feedout pump. The mixing container is primarily used for masses where colouring, flavours, or acids are added, (B) Caramelizing container, (1) Steam infeed, (2) Steam heating (heating coil), (3) Vapour separation chamber, (4) Adjustable valve. Products containing milk are caramelized in the caramelizing container. The degree of caramelization is controlled by the respective level in the container (dwelling time), (C) Vacuumizing container (1) Vapour separation chamber, (2) Flow valve (manually adjustable), (3) Vacuum wheel, (4) Inspection window and lamp, (5) Inspection window, (6) Check valve, (7) Feedout pump. The vacuumizing container cools off the product mass and simultaneously reduces the water content. In a downline mixing container, liquid ingredients are added via dosing pumps. Powder additives are supplied by an auger type filler.

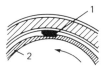

Fig. 16.13   Section of film-cooker (Courtesy of BOSCH—Robert Bosch GmbH, Germany)
(1) Movable scraper, (2) Rotor.

Fig. 16.14   Helical scraper fitted on centre rotor (Courtesy of ter Braak, Netherlands).

By using one, two, three, or four cooking rotors, the capacity of the cooker ranges from 400 up to 1600 kg/h (see Table 16.5). The technical specification of a BOSCH twin-cylinder film-cooker is as follows.

Throughput	approx. 600–800 kg/h (depending on initial dry substance and degree of cooking)
Steam consumption	approx. 200 kg/h
Steam connections	1 in
Supply line	$1\frac{1}{2}$ in
Condensate connections	$\frac{3}{4}$ in
Water connection	$\frac{3}{4}$ in
Connected electric load	approx. 10.5 kVA
Working excess pressure	10 bar
Dimensions	approx. 2800 × 1400 × 2450 mm

Table 16.5   Specifications of continuous film-cookers (rotor cookers) with one, two, or three rotors (Courtesy of ter Braak BV, Netherlands)

Rotorcooker	Single	Twin	Triple	
Working capacity	450	900	1.350	kg/h
Main drive motor(s)	11.5	23	34.5	kW
Power consumption	25	40	58	kVA
Steam consumption	225	450	675	kg/h max.
Steam pressure	10	10	10	bar max.
Air consumption	300	300	300	l/h
Vacuum pump	2.2	2.2	3	kW
Water consumption of vacuum pump at 15°C	800	800	1200 l/h	
Length	3400	3700	3700 mm	
Width	1950	2100	2250 mm	
Height	2900	2900	2900 mm	
Weight	1900	2750	4700 kg	

16.5.3.2   *Operation*

The cylinder cooker is fed with the dissolved sugar solution (about 80% to 85% of DS) from a supply container. In the film cooking section (Fig. 16.12), the revolving rotor scrapers, by means of their centrifugal action and the narrow gap between the scraper and the wall, form a thin film of pre-dissolved solution which is heated and cooked quickly by means of the steam-jacketed wall. With a steam consumption of about 200 kg/h and maximum working excess pressure up to 10 bar, the rotation and the helicoidal action of the Teflon scrapers create high turbulence in the product mass in the annular gap and impel the product towards the outlet of the cooking cylinder. The DS of the product is dependent on the degree of cooking, that is, the dwelling time, which is about 3–5 seconds. With the increase in temperature of the product mass, the moisture is evaporated and drawn off into the vapour separation chamber so that the mass is concentrated to the appropriate dry matter content.

After cooking in the horizontal cylinder and evaporation in the vapour separation chamber, the product mass is transferred into the downline mixing or caramelizing container.

In the mixing container (Fig. 16.12, A), the mass is heated and mixed by the revolving agitator which ensures a uniform mixture as other ingredients such as colourings, flavours, and acids are added. The mixed product is discharged through the feedout pump to a cooling and tempering belt.

In the caramelizing container (B), the layer of product mass close to the inner wall of the container is continuously caramelized and scraped off to prevent burning of the milk particles as the mixing tools blend the newly caramelized product and the bulk mass thoroughly to get a uniform mixture. The degree of caramelization is controlled by the respective level in the container (dwelling time). The caramelized mass is used for hard and soft caramels.

In the vacuumizing container (C), the product mass from the vapour separation chamber through the regulating valve is boiled under a vacuum, as in Fig. 16.10, so that the moisture is further evaporated, which results in a product with a higher DS suitable for chewable candies.

17

Batch former and rope sizer

17.1 SUGAR MASS SUPPLY

After cooking, the sugar mass is generally not supplied directly to the batch former but is mixed with other ingredients, for instance flavouring, acids, fat, or colourings, and cooled to a temperature suitable for batch forming. The mixing and cooling operation can be carried out either on a batch kneader or on a continuous cooling and tempering band which is preceded by a bowl mixer or an in-line mixer.

As shown in Fig. 17.1, the cooked sugar mass is transferred into an in-line mixer (11) by means of the two draw-off rollers (10). During this process the mass itself is used as a sealing agent between the vacuum chamber and the outside atmosphere. After mixing with liquid and powder additives (12) and (13) in the screw mixer (11), the sugar mass flows continuously onto the cooling and tempering belt (14) where it is thoroughly kneaded by the turning ploughs (21) and the kneading rollers (22), and then cooled by means of water jets (19) on the lower side of the steel belt.

For a 10 m cooling belt with an output of 1250 kg/h (Robert Bosch, GmbH), the cold water consumption is 1.5 m³/h at 15°C, and the steam consumption for temperature regulation is 150 kg/layer (sugar mass). The specifications of two other cooling belts are given in Table 17.1.

17.1.1 Cooling

The cooling and tempering unit comprises an endless stainless steel belt running on two rotating drums each of 1000 mm diameter, a framework of sectional steel tubing and water and steam systems. The steel belt is 8 m to 15 m in length and is equipped with a tension device for simple adjustment of the belt tension. Two guide rails fitted under the sides of the belt prevent any lateral deviation. Below the belt, the tempering system is divided into several zones, each of which is independent of the others, and each having its own water circulation system accompanied by a steam heating system. The sugar mass is cooled by the cooled belt which is in turn kept cool by cold water through the spray nozzles (19) against the underside of the stainless steel belt. The temperature of the water is controlled by a thermostat. If the water temperature is too high, some additional cold water is used to cool it down automatically. If the water is

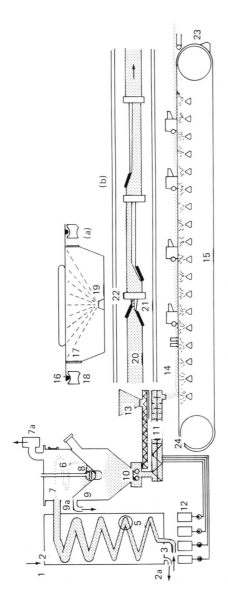

Fig. 17.1   Functional diagram of a cooling and tempering band with a continuous cooker and an in-line mixer (Courtesy of BOSCH — Robert Bosch GmbH, Germany), (a) Section and plan of tempering belt, (b) Plan of kneading process, (1) Steam dome, (2) Steam inlet, (2a) Condensate, (3) Sugar solution, (4) Cooking coil (NiCr), (5) Detail of vortex elements (as in Fig. 16.6(12)), (6) Tangential inlet of the cooked sugar solution into the vapour separation area, (7) Vapour separation area, (8) Valve, (9) Vacuum chamber, (9a) Vacuum line/pump, (10) Draw-off rollers, (11) In-line mixer, (12) Dosing unit, (13) Powder filler, (14) Cooling and tempering belt, (15) Steel belt, (16) Profile strips, (17) Carrier strips, (18) Guide wheel, (19) Tempering unit (water jets), (20) Sugar mass, (21) Turning stations, (22) Kneading station, (23) Washing station, (24) Creasing station.

Table 17.1 Specifications of cooling band installation (Courtesy of ter Braak BV, Netherlands)

Cooling belt	10 metres	15 metres	
Cooling capacity			
hard boiled 130–80°C	300	1.350	kg/h
Main drive motor	1.1	1.1	kW
Power consumption	4.5	5	kVA
Water consumption at 18°C	2500	4000	l/h
Steam consumption	25	25	kg/start up
Length	11 000	16 000	mm
Width	900	900	mm
Height (horizontal)	1600	1600	mm
Weight	3200	4150	kg

too cold, live steam may be employed to heat it up. The operation of the water nozzles is controlled by a flow supervision system.

17.1.2 Kneading

The kneading operation is aimed at accelerating the heat transfer and obtaining a uniformly tempered consistency by means of the turning ploughs and the kneading rollers which are all arranged above the steel belt and cooled by circulating water. To prevent damage to the steel belt, Teflon strips are fitted at the bottom edges of the ploughs. The thickness of the sugar mass is controlled by regulating the centre height of the kneading rollers. To balance the kneading pressure on the belt, counter-pressure rollers are fitted underneath the steel belt at the place where the kneading rollers are positioned.

17.1.3 Cleaning

The kneading rollers can be tilted upwards, which permits them and the steel belt to be cleaned easily and quickly. For food hygiene and to prevent the sugar mass from sticking onto the stainless steel belt, the installation also has a washing station and a lubrication device. The washing station is located underneath the steel band, and consists of a rotary brush and scraper blade which ensure that the steel belt is completely dry and ready to accept the next sugar mass for cooling and tempering. The lubrication device is fitted at the head section of the steel belt; it provides a uniform film of separation grease to separate the adhesive sugar mass from the smooth steel belt.

At the end of the cooling and kneading belt, the plastic sugar mass is fed to the following processing machines, principally the batch former.

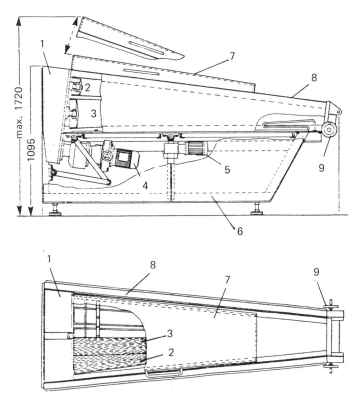

Fig. 17.2   Batch former (Courtesy of BOSCH—Robert Bosch GmbH, Germany), (1) Gear box, (2) Upper roller, (3) Bottom roller, (4) Main motor, (5) Trough lifting and lowering motor, (6) Frame, (7) Trough lid, (8) Trough, (9) Handwheel.

17.2   BATCH FORMER

The batch former shapes the cooled and tempered plastic sugar mass into a filled or unfilled sugar cone for supply to the following rope sizer for a wide range of products such as hard candies, lollipops, milk toffees, and chewable candies. Fig. 17.2 illustrates the batch former. It is generally composed of a trough of adjustable height, housing four conical stainless steel rollers with profiled surfaces, machine frames, and a control cabinet with a facia situated in the right-hand side of the machine. The machine frames cover, of painted mild steel sheet, houses the drives used for actuating the conical rollers, and for lowering or lifting the trough. A different type, the Latini batch former made by Latini Machine Company Inc. U.S.A., is equipped with a motor of 2-HP (1.49 kW) capacity to power the hydraulic system. All operational features such as rotation of the conical rollers, lifting, and lowering, are hydraulically operated. Its capacity is up to 175 pounds (79 kg).

The technical specification for the Bosch batch former is as follows.

Volumetric capacity	75–100 kg
Rope exit	min. 35 mm
Drive motor for conical rollers	0.75 kW, infinitely variable
Drive motor for trough lifting and lowering	0.75 kW
Installed electric heating	220V, 4500 W max.
	3–stage control
Connected electrical load	approx. 7 kVA
Space requirement	2450 × 1090 × 1720 mm
Net weight	820 kg

17.2.1   Trough and conical rollers

The trough of a batch former is commonly mounted on bearings at a pivoting point at the height of the transfer position of the sugar rope, to ensure a smooth delivery to the following rope sizer.

The four rollers are similar in construction and configuration, but with different diameters. They are made of 1 mm stainless steel sheet profiled with evenly distributed raised embossing. The taper is 1:18–20 for the two small rollers, and about 1:10 for the two large ones which are installed at the bottom of the trough, the small ones being fitted at the upper part of the trough.

The four conical rollers (see Figs 17.3, 17.4) are mounted in trapezium fashion, the upper dimension of the trapezium being greater than the lower. The smaller (upper) rollers are centred at the two top corners, and the lower (bottom) rollers at the two bottom corners. The four axes of the rollers are not parallel to one another; they converge, so that the trapezium section becomes smaller and smaller, and so, therefore, does the contained space. The outlet of this device is at the small end, and the product is thence delivered to the next stage, the rope sizer. The inclination of the contained space can be adjusted by handwheel or by a motor via worm gearing which raises or lowers the large end of the trough. The steeper the inclination, the faster the sugar cone runs.

Since the batch size (batch capacity) varies with the variation of the output of the production line and the outlet size of the batch former is dependent on the requirement of the following rope sizer for the rope diameter which changes with the variety of candy formulae, shape and size, the distance between or among the conical rollers should be adjustable to suit different conditions. The handle fitted at the trough large end is generally used to regulate the distance between the two upper rollers. The more the rollers are separated from each other, the larger the batch size allowed. The handwheel at the trough small end (the outlet end) is commonly used for the adjustment of the sugar cone outlet formed by the small ends of the four conical rollers. The closer they are to each other, the thinner the sugar rope that will be fed to the rope sizer.

By means of the rotating rollers, the sugar mass is rolled, squeezed, and formed into a sugar cone as it is turned over and over again, revolving in the opposite direction to the rollers' rotation. As soon as the revolving sugar cone engages with the rope sizer wheels, it will be forced to move linearly without any possibility of rotation. Therefore, if the batch former rollers always rotate in one direction, the sugar cone twisting will certainly take place at the boundary between the batch former and the rope sizer, which would result in stress concentration, abnormal rope sizing, and candy misshaping. Because of this, batch formers are commonly equipped with a reversing device to make the rollers rotate in the same direction, and then are reversed by either a mechanical or electrical device. The mechanical method generally comprises gear wheels, reverse fork handle, and clutch. This makes the transmission complicated. Hence, most batch formers are equipped with an electric reversing device by which the electric motor revolves alternately in the two directions so that the four rollers are made to rotate and reverse, and in turn, the sugar mass is forced to rotate and reverse in the same cycle. The rotation cycle can vary from 1 to 100 seconds, controlled by a timing switch. This feature is especially beneficial for soft or striped batch sugar cone.

17.2.2   Heating

To form the cooled and tempered sugar mass into a smooth sugar cone, it is necessary to keep it in a perfectly plastic condition, that is, in a suitable temperature zone which for the sugar mass from plastic to solid is very narrow. All parts in contact with sugar mass are made of stainless steel which is a good heat conductor, so that the sugar mass would cool quickly, especially in winter. Too low a temperature will cause the product to be dark, rough, cracked, or with other defects, while too high a temperature will cause the sugar mass to be sticky and difficult to handle. To prevent the sugar mass from hardening resulting from overcooling, batch formers are commonly equipped with a heating system, either steam or electric.

The steam heating system consists mainly of steam coils which should be installed under the bottom rollers. That is, the heat energy is transferred to the sugar mass via the bottom rollers by heat conduction, so that the sugar mass tends to stick to the bottom rollers if they are too hot. In this case, the steam pressure is generally lower than 2 kg/cm^2 (2 bar). Furthermore, steam heating is not as convenient as electric heating, so that most batch formers are equipped with electric heaters which are commonly installed within the trough lid. In this way, the sugar cone is directly heated by means of heat radiation from the top electric heater, and the heating efficiency is higher than that of steam heaters.

17.2.3   Sugar cone formation

As shown in Fig. 17.3, the upper and bottom rollers revolve in the same direction. Their profiled surfaces, in contact with the sugar mass, produce enough friction to turn over the sugar mass again and again in an opposite direction to that in which it is rolled and squeezed by the rotating profiled rollers in their trapezium space. The sugar mass is thus gradually formed into a sugar cone and continuously extends and moves toward the batch former outlet by gravity along the slope of the inclined trapezium space. The alternating changes in the direction of revolution of the four rollers make the sugar cone rotate in the same mode (alternately reversing), which permits the sugar cone to enter the rope sizer smoothly without twisting.

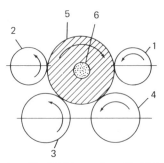

Fig. 17.3   Sugar cone formation, (1), (2) Upper rollers, (3), (4) Lower rollers, (5) Sugar cone,
(6) Centre fitting material.

The heating device, either electric or steam, provides suitable thermal energy
supervised by a temperature control system to keep the sugar mass in a desirable
plastic state for the sugar cone formation to be carried out evenly. The batch forming
technique is widely used in candy production because of its flexibility and for the
advantage that no air is trapped within the sugar mass during the process of cone
formation.

17.2.4   Transmission
For synchronization with other equipment in the production line, the batch former is
generally driven by an infinitely variable motor via gear box and a mechanical reducer
with a large transmission ratio, since the conical rollers are always working at very
low speed, for instance 20 rev/min for the bottom rollers and 33 rev/min for the upper
rollers.

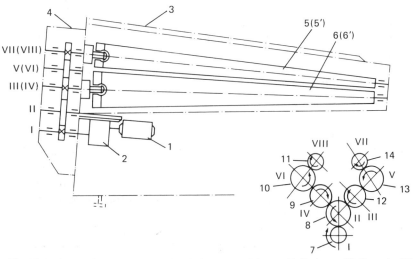

Fig. 17.4   Comparison system of a batch former (1) Motor, (2) Reducer, (3) Trough, (4)
Gearbox, (5), (5′) Upper rollers, (6), (6′) Bottom rollers, (7)–(14) Gears, I–VIII shafts.

As shown in Fig. 17.4, the gear wheel (8) is driven by the motor (1) via reducer (2), shaft I, and gear wheel (7), and transfers the rotation to shaft III and shaft IV via gear wheels (12) and (9) so that the bottom rollers (6) and (6′) are made to rotate in the same direction. Since gear wheels (9) and (12) are also engaged with gear wheels (10) and (13) respectively, which are in turn engaged with gear wheels (11) and (14) respectively, shafts VIII and VII are powered and transfer the rotation to the upper rollers (5) and (5′) to make them rotate in the same direction. Scheme 17.1 shows this transmission expression.

$$\text{Motor 1} \rightarrow \text{Reducer 2} \rightarrow \text{Shaft I} \rightarrow \frac{\text{Gear 7}}{\text{Gear 8}} \rightarrow \begin{cases} \dfrac{\text{Gear 8}}{\text{Gear 9}} \rightarrow \text{Shaft IV} \rightarrow \text{Left bottom roller} \\[2mm] \dfrac{\text{Gear 8}}{\text{Gear 12}} \rightarrow \text{Shaft III} \rightarrow \text{Right bottom roller} \end{cases}$$

$$\frac{\text{Gear 9}}{\text{Gear 10}} \rightarrow \frac{\text{Gear 10}}{\text{Gear 11}} \rightarrow \text{Shaft VIII} \rightarrow \text{Left upper roller}$$

$$\frac{\text{Gear 12}}{\text{Gear 13}} \rightarrow \frac{\text{Gear 13}}{\text{Gear 14}} \rightarrow \text{Shaft VII} \rightarrow \text{Right upper roller}$$

Scheme 17.1   Transmission expression of a batch former.

17.2.5   Centre filling

Candies with centre filling are very popular. The filling materials are introduced into the sugar cone by equipment incorporating Teflon-coated filling pipes of various diameters and lengths, which are dependent on the filling material consistency.

The piston filler and rotary filler (Fig. 17.5) are used for filling liquids and pastes, the augur filler for powders and a fill-extruder for highly viscous materials. The extruder presses the viscous material through the fill pipe, which is accomodated in the incline of the batch former, directly into the sugar cone. In a continuous production line, the fillers can be synchronized with other processing equipment.

17.3   ROPE SIZER

The sugar rope from a batch former is about 35–50 mm in diameter, which is too thick to meet the requirement of the succeeding candy maker (candy shaping machine). A rope sizer, at the conjunction between the batch former and the candy maker, is designed to size and deliver the sugar rope at a suitable diameter to the following candy maker at a variable speed, using four or more flattening rollers in stages.

A rope sizer may be directly fed from a batch former, or via a vertically situated pair of sizing rollers (Fig. 17.6(6)), or via a crossed set of 4 sizing rollers (Fig. 17.7).

A typical rope sizer generally comprises a machine base housing the motor and drive units, a machine table on which the easily accessible sizing roller pairs are fitted, and a control cabinet with an operator panel. The housing for the pair of draw-off

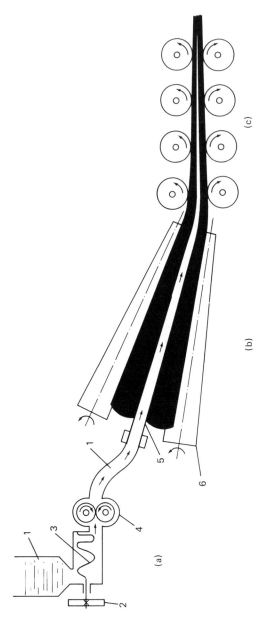

Fig. 17.5   A combination of a centre filler and a batch former, followed by a rope sizer, (a) Centre filler, (b) Batch former, (c) Four-stage rope sizer (Courtesy of Dong Tai City Second Light Machinery Factory, Jiang Su Province, China), (1) Centre filling material, (2) Centre filling pipe, (3) Auger pump, (4) Rotary pump, (5) Centre filling pipe, (6) Bottom roller of batch former.

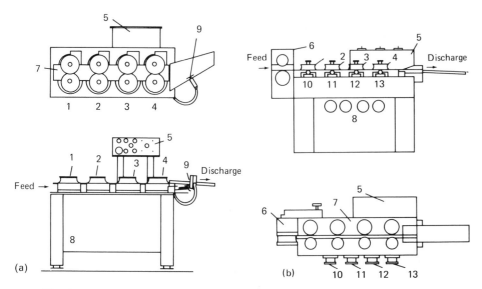

Fig. 17.6   (a) Four-stage rope sizer (Courtesy of Latini Machine Co., Inc., USA, (b) Four-stage rope sizer with independent two-roller pre-sizer (Courtesy of BOSCH—Robert Bosch GmbH, Germany), (1)–(4) Flattening wheels, (5) Control panel with facia, (6) Independent two-roller pre-sizer, (7) Sizing table, (8) Frame housing the drives, (9) Delay tray, (10)–(13) Regulation device on circular section formed by each pair of sizing wheels.

rollers, or alternatively the crossed set of four rollers, is mounted on the machine table as a separate unit.

17.3.1   Sizing rollers

As shown in Fig. 17.6, the flattening wheels are horizontally arranged in line, and their thicknesses are gradually reduced stage by stage with the gradual reduction of the circular section formed by each pair of sizing rollers. To ensure safe pulling of the sugar rope, the side arc surfaces of all rollers are profiled in the shape shown in Fig. 17.7 to increase friction. The sizing wheels are adjustable and their speed can be

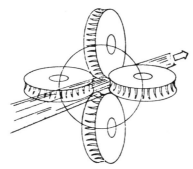

Fig. 17.7   Cross-section of four sizing rollers—pre-sizer (Courtesy of BOSCH—Robert Bosch GmbH, Germany).

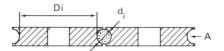

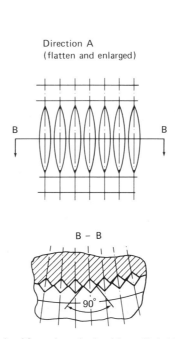

Fig. 17.8   Pair of flattening wheels with profiled side surfaces.

varied to link with the following candy maker. A thermostatically controlled electric heater is positioned beneath the sizing wheels for keeping the sugar rope in the perfect plastic state. For instance, on a Latini four-stage sizer each sizing stage is equipped with a 650 V calrod heater.

Since the diameter of the sugar rope decreases stage by stage, the speed of the sizing rollers should be increased stage by stage. To ensure that the sugar rope is sized and delivered evenly, it is absolutely necessary to maintain constant throughput at every stage, that is

$$Q = A_i V_i = \text{constant},\tag{17.1}$$

where Q is the throughput of sugar rope (cm^3/min), A_i is the circular section (cm^2) of the sugar rope formed by a pair of sizing wheels at stage i, and V_i is the linear velocity (cm/min) of the sizing wheels at stage i.

A_i may be calculated from

$$A_i = \frac{\pi}{4} d_i^2\tag{17.2}$$

where d_i is the diameter (cm) of the sugar rope at sizing stage i, and d_i (Fig. 17.8) is dependent on the radius of the side arc of the flattening wheels at stage i.

$$V_i = \pi D_i n_i \tag{17.3}$$

where D_i is the diameter of the inner cycle of the sizing arc (cm), and n_i is the rotation speed of the sizing wheels at stage i, (rev/min).

Suppose that there are n pairs of sizing wheels, from equation (17.1), there will be

$$d_1^2 D_1 n_1 = d_2^2 D_2 n_2 = \cdots = d_n^2 D_n n_n \tag{17.4}$$

so that

$$\frac{n_2}{n_1} = \frac{d_1^2 D_1}{d_2^2 D_2} \tag{17.5}$$

. . . .

$$\frac{n_n}{n_{n-1}} = \frac{d_{n-1}^2 D_{n-1}}{d_n^2 D_n}. \tag{17.6}$$

Therefore the rotation speed ratio of sizing wheels between any two adjacent stages can be calculated. The result will be used as the basis for the design of the transmission system.

17.3.2 Transmission

The rope sizer is usually driven by an infinitely variable drive motor via a stepless variable belt drive, worm gear box, and gear trains (Fig. 17.9). Bevel gear wheels (5) and (6) transfer the rotation to the flattening wheel A via shaft I on which gear

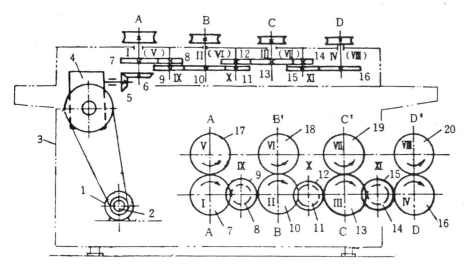

Fig. 17.9   Transmission system of a four-stage rope sizer, (1) Motor, (2) Stepless variable drive, (3) Frame, (4) Worm gear reducer, (5), (6) Bevel gears, (7)–(20) Gear wheels. I–XI shafts, A,A′, B,B′, C,C′, D,D′ sizing wheels.

wheel (7) is engaged with gear wheel (17) on shaft V so that the flattening wheel A′ is powered to rotate in the opposite direction to wheel A. At the same time, gear wheel (7) also actuates gear wheel (8) on shaft IX. By means of gear wheel (9), gear wheels (10) and (18) as well as shafts II and VI, rope sizing wheels B and B′ rotate in opposite directions. In turn, sizing wheels C and C′, as well as D and D′, rotate in opposite directions to each other. Along the ringlike passage formed by the four pairs of rotating wheels, the sugar rope is pulled forward to the following candy shaping machine at the predetermined diameter.

The transmission expression is shown in Scheme 17.2.

$$\text{Motor } 1 \rightarrow \begin{array}{c} \text{Stepless} \\ \text{variable} \\ \text{drive } 2 \end{array} \rightarrow \text{Worm gearing reducer } 4 \rightarrow \frac{\text{Bevel gear } 5}{\text{Bevel gear } 6} \rightarrow \text{Shaft I}$$

$$\text{Shaft I} \begin{cases} \text{Sizing wheel A} \\[4pt] \dfrac{\text{Gear } 7}{\text{Gear } 17} \rightarrow \text{Shaft V} \rightarrow \text{Sizing wheel A}' \\[10pt] \dfrac{\text{Gear } 7}{\text{Gear } 8} \rightarrow \text{Shaft IX} \rightarrow \dfrac{\text{Gear } 9}{\text{Gear } 10} \rightarrow \text{Shaft II} \end{cases}$$

$$\text{Shaft II} \begin{cases} \text{Sizing wheel B} \\[4pt] \dfrac{\text{Gear } 10}{\text{Gear } 18} \rightarrow \text{Shaft VI} \rightarrow \text{Sizing wheel B}' \\[10pt] \dfrac{\text{Gear } 10}{\text{Gear } 11} \rightarrow \text{Shaft X} \rightarrow \dfrac{\text{Gear } 12}{\text{Gear } 13} \rightarrow \text{Shaft III} \end{cases}$$

$$\text{Shaft III} \begin{cases} \text{Sizing wheel C} \\[4pt] \dfrac{\text{Gear } 13}{\text{Gear } 19} \rightarrow \text{Shaft VII} \rightarrow \text{Sizing wheel C}' \\[10pt] \dfrac{\text{Gear } 13}{\text{Gear } 14} \rightarrow \text{Shaft XI} \rightarrow \dfrac{\text{Gear } 15}{\text{Gear } 16} \rightarrow \text{Shaft IV} \end{cases}$$

$$\text{Shaft IV} \begin{cases} \text{Sizing wheel D} \\[4pt] \dfrac{\text{Gear } 16}{\text{Gear } 20} \rightarrow \text{Shaft VIII} \rightarrow \text{Sizing wheel D}' \end{cases}$$

Scheme 17.2. Transmission expression of a rope sizer

17.3.3 Operation and control

The sugar rope from the batch former is about 35–50 mm in diameter. As it is fed to the rope sizer and engaged with pairs of the flattening wheels stage by stage, it is pulled by the profiled surfaces of the sizing wheels and stretched out to become thinner and thinner, defined by the circular sections which are formed by the pairs of flattening wheels at all stages. At the last stage, through the last pair of wheels, the sugar rope is stretched and formed into the preset diameter, for instance 10 mm, which is dependent on the requirement of the following shaping machine.

The circular section formed by a pair of flattening wheels may be adjusted by a handwheel via a worm drive and cam to accommodate variations in candy weight.

The rope sizer may be controlled individually by its own supervision system or by a centre control panel which is usually installed on the rope sizer and which supervises the complete continuous production line including centre filler, batch former, rope sizer, and candy maker.

The rope sizer forms the connecting link between the preceding batch former and the succeeding candy maker, so that the rope sizer should be synchronized with both of them. That is, the input should be equal to the throughput of the sugar rope in unit time. As well as the infinite adaptation of the speed of the individual pairs of sizing wheels, the unit is also equipped with an electrically operated delay tray. When the sugar rope output of the rope sizer is excessive, the delay tray will reduce the speed of the sizing wheels, thus permitting the candy maker to consume the sugar rope. When the excessive sugar rope has been eliminated, the delay tray will give the order to the sizing rollers to resume their normal speed. This feed control is most beneficial for the processing of filled or soft centre sugar ropes (Fig. 17.6(9)).

The specification of a typical rope sizer is as follows (courtesy of BOSCH—.Robert Bosch GmbH, Germany).

Output	Infinitely variable from 7 to 13 m/min sugar rope
Drive	1 variable speed drive 0.37 kW, 1500 rev/min
Heater	220 volt, 1840 W (max.)
Machine dimensions	$1600 \times 710 \times 1100$ mm
Weight (net)	Approx. 620 kg

18

Candy makers

18.1 CANDY MAKER WITH ROTARY DIES

The candy shaping machine, or candy forming machine, often called a candy maker, is normally preceded by a rope sizer and followed by the candy cooling and wrapping equipment.

The most colourful candy is "hard boilings", which is not only rich and varied in formulae (colours and tastes) but also in patterns and formations. Hard boilings, or hard candies, are produced by a hard candy maker with rotary dies which is suitable for any size of production line and is very flexible and easily changeable for variation in the candy varieties. The depositing technique is generally used for large production lines.

18.1.1 General construction

A typical hard candy maker (Fig. 18.1) is a highly efficient automatic machine consisting of a cast frame, sizing unit, rotary die-head (Fig. 18.2), candy discharge belt, sugar fine precipitator, machine drive, and control system.

The frame is a firm, durable cast iron rack which is able to absorb the vibration during candy embossing.

The sizing unit is composed of two interchangeable sizing rollers, mutually adjustable in distance apart by the adjusting handwheel to obtain the desired sugar rope diameter (unit weight).

For safety and food hygiene, the rotary die-head is front shielded. The shield is made of cast aluminium and can be pivoted away pneumatically or by hand.

The machine is usually driven by an infinitely variable motor for synchronizing with the preceding sugar rope sizer or the succeeding candy wrapping machine on the production line, and to obtain an optimum speed over as wide range of production capacity. The machine can be controlled either by a separate control cabinet with a facia or by the supervision centre situated at the sugar rope sizer section.

The forming assemblies fall into two groups, shown in Fig. 18.3, and in Figs 18.4 and 18.5. All of them have a similar rotary die-head, but with a different infeed mechanism for the rope. The former group can be used only for unfilled candy production in which the sugar rope is fed in, separated, and preformed by an infeed

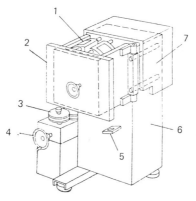

Fig. 18.1 Typical candy maker for hard boilings (Courtesy of BOSCH—Robert Bosch GmbH, Germany), (1) Rotary die-head, (2) Front shield, (3) Sizing roller, (4) Adjusting handwheel, (5) Discharge conveyor, (6) Frame, (7) Gear box.

roller. In the latter group the sugar rope is fed in, preformed, and sectioned by an outer ring which is profiled and eccentrically fitted by means of three location wheels. This group of forming assemblies is suitable for either filled or unfilled candy production.

The forming assembly is also equipped with a sugar-fines scraper to keep the forming die clean, and the scraped sugar fines are collected in a downline filter bag for later reuse, by means of the dust precipitator.

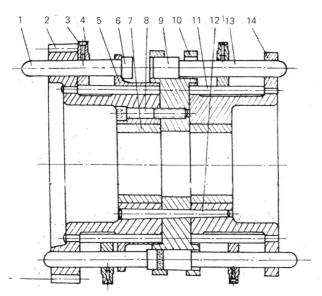

Fig. 18.2 Typical die-head [15] (Courtesy of Food Machinery Factory, Harbin, China), (1) Short die-plunger, (2) Gear drive, (3) Screw, (4) Guide fork, (5) Short die-plunger, (6) End-die of short plunger, (7) Copper sleeve, (8) Bolt, (9) End-die of long plunger, (10) Die-plate, (11) Guide rod, (12) Location pin, (13) Long die-plunger, (14) Supporting plate of long plunger.

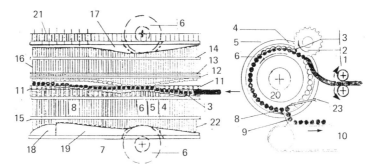

Fig. 18.3   Forming principle of hard boilings (Adapted from the functional diagram of the die-head of the BOSCH model 67B), (1) Rope sizing rollers, (2) Rope infeed roller, (3) Separation and preforming, (4) Insertion, (5) Pre-embossing, (6) Embossing by forming cams or rollers (broken lines), (7) Tension equalization, (8) Ejection, (9) Discharge chute, (10) Conveyer, (11) Die-plate, (12)–(15) Guide plates, (16) Short die-plunger, (17)-(19) Forming cams, (20) Supporting shaft for rotary die-head, (21) Die-head drive gearing, (22) Long die-plunger, (23) Super fines scraper.

18.1.2 Die-head

The die-head (Fig. 18.2) is the centre part of the forming assembly, and is supported on the static centre shaft cantilevered out from the cast iron frame. It generally comprises a die-plate (10), die-plungers (1) and (13) which are guided by the respective guides, that is, by supporting plate (14) and guide forks on the left and right sides. The lantern gear (1) serves to synchronously rotate the die-head. With the rotation, the die-plungers in turn are actuated to strike forward and backward by the cams which are spring-mounted on the cast frame and front shield of the machine. Being subjected to heavy wear, the die-head is made of high grade steel, and the die-plate is

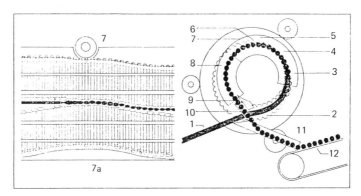

Fig. 18.4   Candy forming by a die-head with a toothed outer ring (Courtesy of BOSCH--Robert Bosch GmbH, Germany), (1) Rope feed, (2) Rope infeed, (3) Preforming, (4) Sectioning, (5) Insertion, (6) Pre-embossing, (7) Embossing by roller and swash plate, (8) Tension equalization, (9) Discharge, (10) Sliding section, (11) Feed chute to distribution belt, (12) Distributor belt.

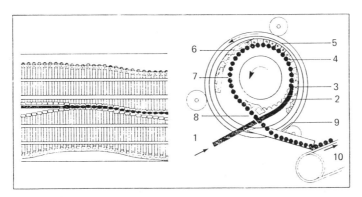

Fig. 18.5   Candy forming by a die-head with an outer ring fitted with blades (Courtesy of BOSCH—Robert Bosch GmbH, Germany), (1) Rope feed, (2) Rope feed-in, (3) Preformer, (4) Embossing die section closed, (5) Embossing, (6) Opening of embossing die section, (7) Ejection, (8) Sliding section, (9) Feed chute to distributor belt, (10) Distributor belt.

surface-hardened. The die-plungers are made by the cold press process, and the forming dies are made of cast copper to allow easy engraving of the die patterns. The semi-spherical bottoms in contact with the forming cams of the die-plungers are quenched to resist the heavy wear in the candy formation process. The guide forks (4) are fitted onto the plunger bodies by means of screws, and the U-shaped (fork) ends ride on the guiding rod (11). In this way, the plungers are kept in stroke axially.

Along the periphery of the die-plate many die holes are evenly arranged in various shapes, for instance round, triangular, square, oval, kidney, which will be the configurations of candy units to be produced. The patterns on the end of the die-plunger (end-dies (6) and (9)) determine the patterns on the candy surfaces. Each hole on the die-plate corresponds to a pair of die-plungers, one long and short. The number of dies on the die-head is closely related to the size (unit weight) of the candy, the more dies there are the smaller the candy will be. An even number is often used, for instance 24, 26, 28, 30, 32, 34, and 36. The candy volume is preset in the separation process. However, the longer the die-plunger stroke is, the thinner the candy unit will be (or the less the candy height will be). Conversely, the thicker the preset candy height is, the shorter the die-plunger stroke should be. In addition, the thicker the sugar rope is, the greater the unit volume of candy will be.

For different varieties of candy, the die-head is the only part which has to be changed. This may be carried out by opening the front shield, dismounting the die-head, and then assembling the new one. The opening and locking of the front shield can be accomplished automatically or by hand.

18.1.3   Candy shaping principle

Candy shaping (forming) is a continuous automatic multi-position process. In Fig. 18.3, the unfilled sugar rope coming from the rope sizer is transferred by a pair of sizing rollers (1) to the rope infeed section of the candy maker. The size of the circular section formed by the sizing rollers is closely related to the unit weight of the final candy products. The toothed feed-in roller (2) is fitted above the outer channel of the die-head. It squeezes the sugar rope into the separation and preforming section (3) ahead of the long die-plunger (22). Here a piece of sugar is cut off from the sugar rope and preformed into a pillow shaped unit by means of the coordinated infeed roller and the toothed ring of the die-plate. Driven by the forming cams (17), (18), and (19) and others fitted on the support frame and the front shield of the machine, each pair of die-plungers strikes forward and backward, that is, they reciprocate face to face or back to back as the die-head rotates, powered by the gear drive (21). In the forward stroke of the die-plungers, the newly cut (separated) sugar unit is pushed by the long die-plunger (22) into the die hole of the die-plate (11) where the long die-plunger (22) and the short die-plunger (16) both strike forward so that the pillow-shaped sugar unit is pressed and embossed into a die-shaped candy product. The candy shaping (embossing) is gradually performed by means of the die-plungers in turn as they are pushed by the curved surfaces of the cams. The heaviest embossing action takes place at point (6). The latest types of candy maker employ embossing rollers (dotted lines) to change sliding into rolling friction.

The embossed candy unit in the die-plate passes the tension equalization section (7) and arrives at the ejection section where the raised curve of the cam (19) pushes the die-plunger a further stroke forward. Then the candy unit is pushed out of the die-plate (11) and discharged through the chute (9) onto the transport conveyor (belt) (10) which leads to the candy cooling equipment. If the candy piece is cold enough as it is ejected at point (8), the discharge chute may be eliminated. During candy shaping some sugar fines tend to stick on the die-plate, which could cause trouble in the subsequent candy shaping process. To prevent this, the scraper (23) is usually employed for sugar-fine cleaning, and a dust precipitator is used for collecting the scraped sugar fines.

As mentioned earlier, this candy forming mechanism is suitable only for the production of unfilled candy products, not because of the die-head but because of the infeed mechanism by which the sugar unit is separated from the sugar rope in only two or three steps by the small infeed roller and the coordinated toothed ring of the die-plate. A filled sugar rope would not be completely closed at the newly cut (separated) end, and the filling materials are often a viscous liquid or paste (jam) or powder which would cause trouble in the candy forming process.

For filled candy products, the rope infeed roller is replaced by either a static or a rotary outer ring which is eccentrically located by three wheels. Coordinated by the toothed inner ring of die-plate, the profiled outer ring gradually engages with the filled sugar rope and separates it into pieces after many steps forward to the centre, so that the filled material is sealed inside the individual candy unit without leakage.

In Fig. 18.4, the die-head operates with the eccentrically rotating toothed outer ring and the inner die-plate ring. The die-plungers are powered by a separating cam-roller and an inclined idling swash plate. After gradually separating from the sugar rope, the sugar unit with sealed ends is pushed into the die-plate by die-plungers on both sides. The embossing process which follows is similar to that shown in Fig. 18.3.

In Fig. 18.5, the outer ring is fitted with several blades which separate the sugar unit from the infeed sugar rope and preform it into a pillow shaped piece. The spring mounted rotary outer ring presses the two embossing sections (the long die-plunger and short die-plunger sections), with a lever-bearing in the base support. Since cam-rollers are employed in the embossing process, the noise is reduced to approximately 83 dBA, The rollers are run in enclosed cams. Furthermore, the long dwelling time guarantees the formation of a firm sugar cover.

Of course, the mechanisms shown in Figs 18.4 and 18.5 are also suitable for the production of unfilled products as well as filled ones. As the candy diameter is 14 to 18 mm, the filling percentage will be 18 to 20 for liquid, 22 to 25 for fruit, 22 to 32 for fat, and 32 to 43 for paste. The specification of a typical candy maker for hard boilings is as follows (Courtesy of BOSCH—Robert Bosch GmbH, Germany).

Throughput	up to 125 m/min, rope speed depending on the type of forming die,
Candy size	Length 27 mm, weight 4.8 g = approx. 1200 kg/h or approx. 4300 candies/min.
Heating	220 V, 150 W (max.)
Cooling	2 blowers, 0.6 kW
Sugar fines suction	1 extraction unit, 3.0 kW
Machine dimensions	2600 × 1175 × 1565 mm
Weight	approx. 1300 kg
kVA (standard)	8 kVA

18.1.4 Transmission

As shown in Fig. 18.6, the machine is driven by an infinitely variable electric motor via belt wheels (2) and (3), clutch (5), and gear wheel (6) which activates the lantern gear (7) on the die-head (16) so that the die-head rotates. Simultaneously, gear wheel (7) engages with gear wheel (8) by which gear wheel (9), and in turn gear wheels (10), (11), and (12), rotate so that the infeed sizing rollers rotate in opposite directions.

The arrangement of this transmission system is simple, compact, and efficient. However, it requires great mechanical precision. The position of the gear wheel shaft, and its bearings, and the relative positions of all the shafts, must be well defined and not liable to variation during use. This design is suitable for machines with moderate loading, a short cantilevered shaft, a medium speed, and a not too complex mode of action.

18.1.5 Operation and control

18.1.5.1 *Adjustment of candy thickness*

The candy thickness, 'candy height,' is closely related to the stroke travel of the die-plungers, which is in turn dependent on the position of the spring-mounted forming cams, so that the thickness of candy can be regulated to some extent by changing the position of the top point of the forming cams. This operation may be carried out by first loosening the lock nut of the regulation handle, then turning the handle to push the forming cams (17) and (19) on rollers (broken lines, Fig. 18.3) forward or backward. If the top point (6) is raised, the candy unit will be thinner and more

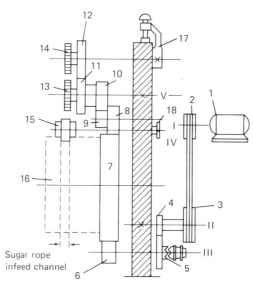

Fig. 18.6   Transmissions system of candy maker (Courtesy of Food Machinery Factory, Harbin, China), (1) Motor, (2), (3) Belt wheels, (4) Gear wheel, (5) Clutch, with a gearwheel, (6)–(12) Gear wheels, (13), (14) Sugar rope sizing rollers, (15) Infeed roller, (16) Die-head, (17) Adjusting handle of rope sizing roller, (18) Adjusting handwheel of infeed roller pressure, I–VI shafts.

compact in texture. Conversely, if point (6) is lowered, the candy will be thicker and its texture will be looser. This regulation should be carried out in association with the sugar rope feed velocity and separation roller or the outer ring, until candy products with smooth surfaces and complete edges and corners are produced.

18.1.5.2   Control of the gap between the die-plunger and die-plate during embossing
The size of the gap is closely related to the quality of the candy formation (embossing). Too small a clearance will cause severe friction between the die-plunger and die-plate, and it will be difficult to push the candy piece out of the die, while too great a clearance results in more sugar fines and deformed candy pieces with extended tails (edges). Different die materials have different heat expansions. The temperature suitable for candy formation is in the range 80 to 90°C. The control of the clearance is generally in accordance with the sliding effect of the plunger after assembly, that is, the assembled unit should permit the plunger to slide freely without any gripping or blocking, even if the temperature of the die-plunger and the die-plate reaches 80°C. Under these preconditions, the clearance distance between the die-plate and die-plungers is commonly about 0.2 mm.

18.2   CANDY MOULD DEPOSITING MACHINES

18.2.1   Introduction
The candy mould depositing technique was developed by Baker Perkins (now APV Baker) twenty years ago. Differing from the traditional candy making process,

described in Chapter 17 and Section 18.1, the depositing process permits high boiled or low boiled sugar to be fed directly to the hopper of the candy depositor (1) (Fig. 18.7), which accurately deposits the cooked sugar into the individual candy moulds. The moulds are directly or indirectly fixed to the chains of the mould conveyor (2) which carries the deposited moulds through a cooling tunnel (3) to the candy ejector (4). The candy pieces formed are ejected on to the candy discharge conveyor (5) which may directly feed them to a high-speed wrapper or an enrobing machine. The emptied moulds may be sprayed with edible oil in every circuit or at intervals to suit the product. For some products the moulds may not need to be sprayed.

18.2.2 Depositing head and its function
The candy depositing head is the key component of the continuous depositor (Fig. 18.8). It can be used for hard candies, fondant creams, toffees, jellies, or gums. The machine frame houses the transmission gears and the drive motor, whose speed is variable for synchronization of the continuous mould movement and other equipment in the production line.

The depositing head generally consists of one or two hoppers with heating and adjusting facilities, and piston cylinder pumps.

The hoppers are made of stainless steel and are used for cooked sugar and centre filling material in filled candy production or for different coloured and flavoured sugars in layered candy production. The hopper sides are fitted with electric heating jackets which are designed in two halves and are held in place by spring clips, permitting a convenient assembly and maintenance. The hopper base, in which the depositing pumps are housed, is also electrically heated. The heated base and side jackets have independent variable heat input controls which may be duplicated for the two hoppers used in centre filling or layered candy production.

The hopper may be heated by oil circulating within the hopper side jackets instead of an electric heating system. In either case, the temperature indicating control instruments, together with the machine supervision system, are housed in a free standing floor mounted control console (see Fig. 18.14, later).

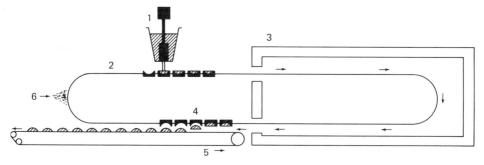

Fig. 18.7 Candy depositing process (Courtesy of APV Baker, UK), (1) Depositor, (2) Mould conveyor, (3) Cooling tunnel, (4) Candy ejector, (5) Discharge conveyor, (6) Mould sprayer.

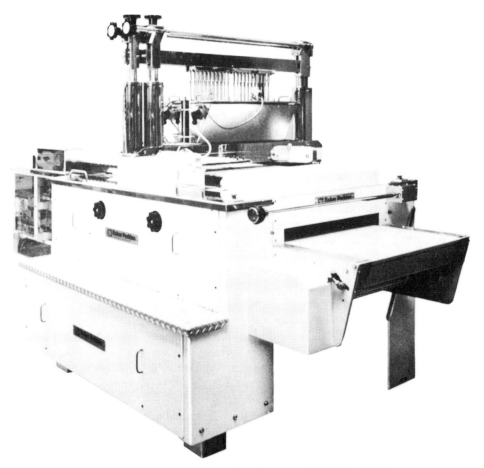

Fig. 18.8   Typical candy depositor (Courtesy of APV Baker, UK).

A set of stainless pistons corresponding to the set of copper cylinders are fitted inside the hopper which supplies the cooked sugar to the cylinders for the pistons to extrude measured deposits through nozzles (Fig. 18.9(b)) in to the individual moulds (impressions) beneath. The weight of the deposit is controlled by the piston travel which can be changed by a handwheel with an engraved scale. The longer the travel, the heavier the candy will be.

The depositor pumps may be in a single or double row, depending on the output required. The pump cycle/min is shown in Tables 18.1 and 18.2. At the end of each stroke, the pump chamber is closed by a patented ball valve within the nozzle, which has a suction effect and prevents sugar tailing and gives accurate deposits.

For centre filling candies, a special nozzle assembly brings the two materials into a concentric nozzle where they are deposited simultaneously (Fig. 18.7 and Fig. 18.11, later).

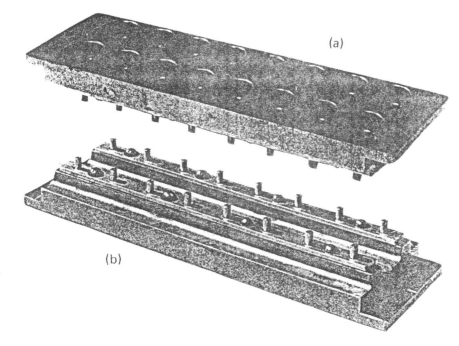

Fig. 18.9   Candy depositing mould (a) Mould, (b) Nozzle.

18.2.3   Depositing moulds

For hard candies such as clear fruits and mints, butterscotch, barley sugar, and milk boils, the moulds are often high pressure aluminium castings coated with PTFE (Fig. 18.9(a)). Each mould is fixed at a certain pitch (76 mm for APV Baker) on the chain of the circuit conveyor. Each mould impression is fitted with a spring loaded ejection pin which is PTFE sleeved and set flush with the mould bottom to eject the sweets out of the impressions as they are depressed by a mechanically operated lever at the ejection point in the return run of the conveyor.

For low boiling sweets, including toffees, caramels, fudge, fondant creams, and certain quick setting jellies, the moulds are made of either a special hard-wearing rubber with a high slip characteristic so that generally no lubrication is needed, or high pressure cast aluminium which should be sprayed with edible oil at intervals to suit the application. The individual moulds can be directly fixed to the conveyor chain at a certain pitch or indirectly fixed to the conveyor chain by means of a common carrier frame. Compared to the traditional starch mould used in the production of jelly and fondant centres which are later enrobed, the starchless moulders, including the flexible and rigid moulds of the depositor, will produce candy sweets or fondant centres ready for enrobing in less than one hour of starting the production line, while with the starch mould it can often be up to twenty-four hours before the centre is suitable for enrobing.

Table 18.1   Specifications of depositing line for soft candies (Courtesy of APV Baker, UK)

		Soft candy–10 minutes cooling											
	Plant size	250			600				950		1300		
	Moulds	091	101	111	181	201	221	271	301	331	361	401	441
	Max. deposit (grams)	9.5	8	6	9.5	8	6	9.5	8	6	9.5	8	6
Cooler size													
8	Units/min	351	390	429	702	780	858	1053	1170	1287	1404	1560	1716
	Pump cycles/min	← 39 single →											
	No. of moulds	474	474	474	948	948	948	1422	1422	1422	1896	1896	1896
9	Units/min	387	430	473	774	860	946	1161	1290	1419	1548	1720	1892
	Pump cycles/min	← 43 single →											
	No. of moulds	523	523	523	1046	1046	1046	1569	1569	1569	2092	2092	2092
10	Units/min	423	470	517	846	940	1034	1269	1410	1551	1692	1880	2068
	Pump cycles/min	← 47 single →											
	No. of moulds	572	572	572	1144	1144	1144	1716	1716	1716	2288	2288	2288
11	Units/min	468	520	572	936	1040	1144	1404	1560	1716	1872	2080	2288
	Pump cycles/min	← 52 single →											
	No. of moulds	621	621	621	1242	1242	1242	1863	1863	1863	2484	2484	2484
12	Units/min	513	570	627	1026	1140	1254	1539	1710	1881	2052	2280	2508
	Pump cycles/min	← 57 single →											
	No. of moulds	664	664	664	1328	1328	1328	1992	1992	1992	2656	2656	2656
13	Units/min	549	610	671	1098	1220	1342	1647	1830	2013	2196	2440	2684
	Pump cycles/min	← 61 single →											
	No. of moulds	713	713	713	1426	1426	1426	2139	2139	2139	2852	2852	2852

Note: Numbers of moulds are based on 2″ pitch
Centrefill applications are limited to 55 strokes/min.

The impressions can be any shape suitable for the depositing process. Fig. 18.10 is a sectional diagram of an individual depositing mould impression, and Table 18.3 gives data for some candy moulds. The minimum angle on metal moulds is 5° for successful demoulding. For centre filling, the depth Y (Fig. 18.10) is recommended by the manufacturer (APV Baker) to be not less than 9 mm without 'a reduction in the percentage of filling being used and certainly no less than 8 mm with any centre material'.

The moulds are lifted at the depositing point by a pneumatically operated device to eliminate tailing of the more viscous materials. The continuous mould movement is in synchronization with the depositing head reciprocating in a horizontal plane.

18.2.4   Top moulding unit

The hard candy can be shaped with or without a top dimple. Top-dimpled products are made by a top moulding unit which is in a section additional to the normal arrangement. After the liquid candy syrup has been deposited, the male moulds of the

Table 18.2   Specifications of depositing line for hard candies (Courtesy of APV Baker, UK)

Solid hard candy

Plant size		250			600			950			1300		
	Mould	113	102	092	233	202	182	333	302	272	443	402	362
Cooler size	Max. deposits (grams)	6	8	9.5	6	8	9.5	6	8	9.5	6	8	9.5
4	Units/min.	770	465	418	1540	930	836	1980	1395	1254	2640	1860	1672
	Pump cycles/min.	70S	46.5S	46.5S	70S	46.5S	46.5S	60S	46.5S	46.5S	30S	45.5S	45.5S
	No. of moulds	186	186	186	372	372	372	558	558	558	744	744	744
5	Units/min.	946	580	522	1892	1160	1044	2838	1740	1566	3784	2320	2088
	Pump cycles/min.	43D	58S	58S	43D	58S	58S	43D	58S	58S	43D	89D	29D
	No. of moulds	218	218	218	436	436	436	654	654	654	872	872	872
6	Units/min.	1144	700	630	2288	1400	1260	3432	2100	1890	4576	2800	2520
	Pump cycles/min.	52D	70S	70S	52D	70S	70S	520	35D	35D	52D	35D	35D
	No. of moulds	250	250	250	500	500	500	750	750	750	1000	1000	1000
7	Units/min.	1320	800	720	2640	1600	1440	3630	2400	2160	4840	3200	2800
	Pump cycles/min.	60D	40D	40D	60D	40D	40D	55D	40D	40D	55D	40D	40D
	No. of moulds	282	282	282	564	564	564	864	864	846	1128	1128	1128
8	Units/min.		920	828		1840	1656		2760	2484		3680	3312
	Pump cycles/min.		46D	46D		46D	46D		46D	46D		46D	46D
	No. of moulds		314	314		628	628		942	942		1256	1256
9	Units/min.		1040	936		2080	1872		3120	2808		4160	3744
	Pump cycles/min.		52D	52D		52D	52D		52D	52D		52D	52D
	No. of moulds		346	346		692	692		1038	1038		1384	1384
10	Units/min.		1160	1044		2320	2088		3300	2970		4400	3960
	Pump cycles/min.		58D	58D		58D	58D		55D	55D		55D	55D
	No. of moulds		378	378		756	756		1134	1134		1512	1512
11	Units/min.		1200	1080		2400	2160						
	Pump cycles/min.		60D	60D		60D	60D						
	No. of moulds		410	410		820	820						

Number of Moulds based on 3"pitch. The letter 'S' or 'D' following the pump speeds denote a single or double row deposit.

Centre filled

		Type 250F			Type 600F			Type 950F			Type 1300F		
Cooler size	Moulds	113	102	092	233	202	182	333	302	272	443	402	362
3	Units/min.	572	350	315	1144	700	630	1716	1050	945	2288	1400	1260
	Pump cycles/min.	52S	35S	35S	52S	35S	35S	52S	35S	35S	52S	35S	35S
	No. of moulds	154	154	154	308	308	308	462	462	462	616	616	616
4	Units/min.	605	460	414	1210	920	818	1815	1380	1242	2420	1840	1656
	Pump cycles/min.	55S	46S	46S	55S	46S	46S	55S	46S	46S	55S	46S	46S
	No. of moulds	186	186	186	372	372	372	558	558	558	744	744	744
5	Units/min.		550	495		1100	990		1650	1485		2200	1980
	Pump cycles/min.		55S	55S		55S	55S		55S	55S		55S	55S
	No. of moulds		218	218		436	436		654	654		872	872

Table 18.3   Data for some metal moulds for hard candies (Courtesy of APV Baker, UK)

Sweet shape	Maximum piece size	Mould pattern 9 × 2	Mould pattern 10 × 2	Mould pattern 11 × 2	Mould pattern 11 × 3
Square with dimple	17.44 sq × 11.5 dp	possible	exists	possible	possible
	18.5 sq × 10.5 dp	possible	exists	possible	possible
	19.5 sq × 10.2 dp	possible	exists	possible	possible
Square† without dimple	18.97 sq × 8.5 dp	possible	possible	possible	exists
	18.89 sq × 10.5 dp	possible	possible	possible	exists
	19.5 sq × 9.5 dp	possible	exists	possible	not possible
	20 sq × 9.5 dp	possible	exists	possible	not possible

†Indicates suitability for centre filling at maximum piece size. Mould pattern 9 × 2 has 18 impressions arranged in two rows of nine.

top moulding unit immediately enter the normal mould so that the two moulds are held together for about one minute. Since the top moulding unit is designed with a separate circuit superimposed over the normal mould circuit between the depositing head and the cooler, the candy can be formed and cooled sufficiently in one minute. This allows the male mould to be then withdrawn without damaging the candy shape, as the ejector pins pass under an ejector shoe and are depressed by 1 mm to eliminate any sticking of the candy pieces to the male mould of the top moulding unit. The conveyor is continuous, and the moulds are cooled on their return to the start position.

18.2.5   Depositing mould circuit and cooling tunnel
Having received the measured deposits, the moulds carried on the conveyor chains travel the length of the cooling tunnel and return upside down to the ejection point (see Fig. 18.14, later).

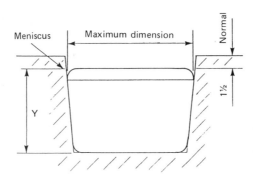

Fig. 18.10   Cross-section of an individual depositing mould impression (Courtesy of APV Baker, UK).

The mould conveyor circuit is driven by a variable speed drive housed within the depositor. The conveyor chains are fitted with quick release spring studs which allow each mould to be released instantly so that a mould can be removed and replaced without interruption of production. Furthermore, the hinged removable covers of the cooling tunnel allow complete access to the mould circuit for maintenance.

For hard candies, cooling is carried out by using conditioned ambient air (not exceeding 21°C, 60% R.H.) for about five minutes.

For low boilings, the cooling tunnel is commonly insulated and equipped with refrigeration compressors, refrigeration evaporation coils, and fan(s) to circulate the air. The evaporation coil, including the four sections, is controlled by a thermostatic controller which can automatically switch in or out the coil sections to suit the cooling load. Softer candies and low boilings such as caramels, fudges, jellies, and fondant creams require about 8 to 10 minutes cooling/setting time.

18.2.6 Candy ejection and discharge conveyor
After cooling, the candy has solidified and is firm enough to be ejected from the mould. The ejection mechanism linked to the depositing head is located at the return travel of the mould conveyor circuit next to the depositing head, and it swings with the mould movement. The inverted mould runs underneath the ejection mechanism.

For low boilings, the ejector is operated by compressed air to make the candy products fall on to the discharge conveyor band (Fig. 18.11). The compressed air requirement is up to 40 m³/hour at 6 bar (APV Baker).

For hard candies, the individual pieces are commonly ejected from the mould by plastic or metal ejecting pins which are spring loaded and depressed by mechanically operated levers.

The ejected candy products are conveyed from the ejection point to the wrapping or enrobing machine by the discharge conveyor. This is plastic coated and fitted underneath the moulding circuit. Along the length of the discharge conveyor, tracking rollers and a supporting band are usually fitted to ensure smooth operation.

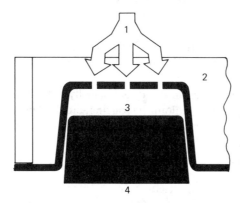

Fig. 18.11   Candy ejection by compressed air (Courtesy of APV Baker, UK), (1) Compressed air, (2) Moulder, (3) Depositing impression, (4) Ejected product.

18.2.7   Operation

18.2.7.1   Cooked sugar supply
The candy depositing process is a highly efficient continuous operation. The candy depositor should be fed from a continuous cooking equipment, as shown in Fig. 18.14. Where only batch cookers are available, an intermediate reservoir should be installed, from which the cooked sugar liquid can be continuously fed to the depositor hopper. In this case, attention should be paid to toffee production. It is suggested that many small batches be made rather than large ones. Otherwise, variations in product colour and taste are difficult to avoid.

For hard candy continuous cooking and depositing, the total reducing sugar (on a dry solid basis) should be kept within the limits of 16% to 22%. Below 16%, the products will tend to exhibit general graining. To avoid picking up moisture, the hard candies should be wrapped while still warm and in an air conditioned environment, say 40% R.H., since the relative vapour pressure of the candy under these conditions is higher than that of the surrounding atmosphere.

18.2.7.2   Centre filling
Jam, fat mixtures, and some liquids are often used as centre filling materials for which the viscosity and solids content are very important to give a good candy centre shape and successful wrapping.

The viscosity of the centre filling material should be kept as close as possible to that of the casing sugar by changing the temperature of the material or by modifying the ingredients.

The solids content of the centre filling should be as high as possible to lower the water migration rate from the wetter centre to the outside, resulting in crystallization. This also reduces the vapour bubbles which weaken the candy shell since the water in the centre filling material can boil as it comes into contact with the hot casing sugar. Therefore, appropriate care is often needed. The jam is generally made from fruit pulp, glucose, sugar, and a small amount of water. The viscosity is mainly dependent on the type of fruit pulp and its proportion, so that if it is too thick, resulting from a jam with a high pectin content, it may be buffered with a suitable salt. If the jam is too fluid, the proportion of fruit pulp should be increased.

The fat mixture is usually made from fat and icing sugar, or chocolate, or compound chocolate. This should be melted and kept at a suitable temperature (say 40°C) for depositing.

The liquid material for centre filling is basically a mixture of cooked sugar and glucose mixed with glycerine, colour, flavouring, etc. In this case, a lower percentage of filling than normal is often necessary to ensure smooth operation and to prevent any leakage of the centre filling.

The centre filling process is illustrated in Fig. 18.12.

18.2.8   Output
For fondant and jelly production, the moulding depositor does away with starch and provides a clean, hygienic, and space-saving technique for manufacture. It also has

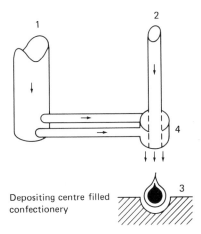

Fig. 18.12   Centre filling process (Courtesy of APV Baker, UK), (1) Casing hopper, (2) Centre filling hopper, (3) Mould, (4) Compound nozzle.

advantages in the extremely short working cycle compared with the starch mould process.

For hard candies, the mould depositor provides products with a glass-like appearance as the cooked sugar is deposited in the molten state and cooled until solid. That is, the finished products are smooth and transparent except when milk or butter are added. The smoothness and clarity of the sweets give them extra marketing appeal.

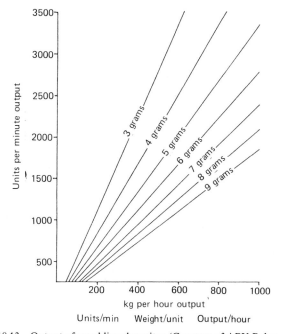

Fig. 18.13   Output of moulding depositor (Courtesy of APV Baker, UK).

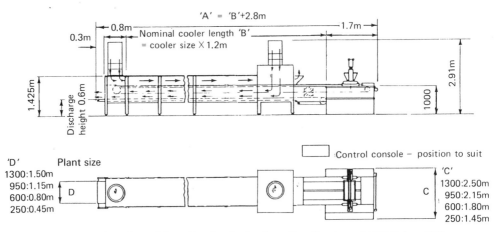

Fig. 18.14   Dimensions of depositing line for hard candies (Courtesy of APV Baker, UK).
(Plant size relates to Tables 18.1–18.3)

Furthermore, depositing produces a variety of sweets with accurate shapes and weights, but requires less labour (one operator can supervise two production lines), thus giving good cost control and high efficiency.

The output of a moulding depositor is normally expressed in units of candy per minute, as shown in Fig. 18.13 and Table 18.1 as a general guide. It is dependent on the candy unit weight and shape as well as the cooling time required. The cooling time depends on the candy recipe, the type of sweet, and the proportions of the product shape. For instance, hard candies may require five or more minutes, fondant creams and toffees about eight minutes, and other sweets, such as jellies, may need ten minutes to cool. On the other hand, the thinner the product is, the shorter the cooling time required.

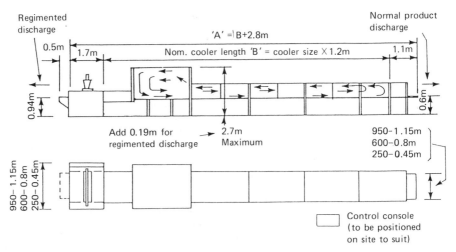

Fig. 18.15   Dimensions of depositing line for soft candies (Courtesy of APV Baker, UK).

The output given in Table 18.2 is based on a minimum cooling time of five minutes for hard candies and is dependent on the final moisture content of the product not being more than 3.5% (Karl Fischer) with the ambient cooling air not exceeding 21°C and 60% R.H. The letters 'S' and 'D' following the pump speeds denote a single or double row deposit. Fig. 18.14 shows the dimensions of this production line.

The data shown in Table 18.1 are based on a typical 10 minute cooling cycle for soft candies. If the product can be cooled and successfully ejected in a shorter time, say 9 minutes, then the cooler can be reduced in length to $\frac{9}{10}$ of that shown in Table 18.1. That is, a size 10 cooler would be reduced to size 9. Alternatively, the machine could cycle faster to give an increased output. Of course, the speed of the cooling circuit is limited by the speed of the depositing pump. Fig. 18.15 shows the dimensions of this production line.

Part IX

Food machinery design

19

Food machinery design

The design of food machinery is a rather complex subject which not only needs all the knowledge necessary for common machinery design but also calls upon food engineering and technology, food science, food rheology, food hygiene, and machinery moulding arts.

19.1 GENERAL REQUIREMENTS

A food machine should be designed to present a clean external appearance. Food quality paints, especially light coloured, are commonly used as the machine body finish, while some machines are also decorated with stainless strips and trims.

The products made by food machinery are supplied to the consumer. Some of the products are directly eaten, so that food hygiene is vitally important. Therefore, all parts in contact with food materials should be made of non-toxic and corrosion-resistant materials. It would be reasonable to keep these parts and the transmission system separate from one another, with proper sealing arrangements, so that no lubrication oil can leak out and pollute the food materials or products.

All parts in contact with food materials should be designed to be easily cleaned, and some of them should be designed to be easily assembled and disassembled for washing. For instance, the feed screw assemblies for the dough and filling materials of the dumpling maker may be disassembled in one to two minutes for thorough cleaning. If necessary, some machine constructions can be designed to allow the rinse water to be discharged smoothly without accumulated water or dregs getting into the transmission system to affect machine operation.

Working safety is of the first importance. Machines should be designed for easy operation and maintenance. All locations which could give rise to accidents from negligence should be equipped with protective covers, danger signs, and other safety precautions.

19.2   BASIC STEPS OF FOOD MACHINERY DESIGN

Food processing machines vary widely in configuration, size, and material construction. However, their design procedures are much the same; that is, design preparation, planning design, technical design, and evaluation.

19.2.1   Preparation of food machinery design

In this stage, a close study of the assignment document for the food machine to be designed should be carried out, since it is the basis of all the following design procedures. The object and requirements of the design must be thoroughly examined, and then the amount of design work should be estimated. The engineers responsibile for the machine design should investigate thoroughly the realities of the food production siting, and should be familiar with the food materials to be processed, the end products, the technological conditions, the operating conditions, etc. They must read the technical literature, collect the relevant data, and try to find out about recent developments in the appropriate science and technology.

19.2.2   Planning design

This stage is based on a full study and investigation of the assignment document. Several different plans should be formulated and compared, according to different technological processes, from which the most reasonable plan can be finally chosen.

Planning design consists of the selection of technological process, determination of key technical data, overall plan drawings, sketches of main subassemblies, transmission diagrams, sketches of control and working cycles, and reports of experimental, technical, and economic analyses.

For analysis and reference, simulation and material tests are often needed to get reliable data. Only when desirable results are achieved, can they be applied to the actual design.

After the planning design has been worked out, it should be submitted to a committee of experts and executives for examination and approval.

The approved planning design will then be the basis for the rest of the exercise.

19.2.3   Technical design

Technical design generally consists of drawings of the complete machines of subassemblies, and of individual elements, machine design and operation instructions, and various tables, which will be presented complete to the machine manufacturer as the basis of his activities.

19.2.4   Evaluation

Having done all the design work, the machine designers should make a summary of the characteristics, the level of science and technology, and the possible demerits of the designed machine, so that an improved design could be offered later. All the documents of the design should be put in order and filed.

19.3   TECHNICAL DESIGN

After the overall design, it is necessary to carry out the technical design so that the manufacturer can start thinking about machining, assembling, and regularizing according to the technical documents (drawings, design instructions, and tables) provided by the technical design.

Technical design forms a connecting link between the preceding overall design and the succeeding manufacture. It must not flout the main principles settled in the overall design. It is a vital step in food machinery design, and it should be carried out seriously and carefully.

19.3.1   Construction design

The overall design is the general plan of the machine, in which the structure and arrangement of the drive and operating mechanisms are designed from the overall point of view, but not necessarily in great detail. In technical design, it is imperative to have the dimensions, configurations, materials, and relative positions of all parts and assemblies of every mechanism for further consideration. At this stage, partial variations may be allowed but they should not affect the preset overall design.

A sketch showing the elements' dimensions, configurations, and relative positions is always presented.

19.3.2   Calculations

The dimensions and configurations of the elements defined in the construction design should be calculated and checked to see if they can meet the requirements of strength and rigidity. For the driving elements, the transmission ratio, anti-destructive capability, and load-bearing capacity should also be calculated and checked. For some elements, calculations on kinetics and dynamics are also needed. If necessary, amendments can be made.

19.3.3   Drawings

Calculation is the preparation for the technical design, which is followed by the preparation of drawings. This is the most labour intensive work in the whole process. Drawings, including overall assembly drawings, subassembly drawings, and element drawings, are part of the formal documents to be presented to the manufacturer as the basis for machining and assembling. Any mistake in the drawings will result in an economic loss, so that especially vigilant attention should be exercised at this stage.

The drawing office work generally starts with overall assembly drawings followed by subassembly drawings, and ends with individual element drawings, of which the inner parts are preferably drawn before the outer parts.

19.3.4   Instructions

These documents, which should be provided along with the drawings, are the instructions for the machine design and operation.

The design instructions generally comprise design basis, technical functions, economic analyses, demonstrations of overall plan, and all analyses and calculations on transmission systems, the movement of the mechanisms concerned, strengths and rigidities of main parts, exposition on the merits and demerits, and the necessary explanations of the design, etc.

The operation instructions are composed of the application to a suitable range of food products, sketches of working principles, construction and transmission; instructions about shipping, case-opening, and installations, instructions on regulations, lubrication and maintenance, possible trouble diagnosis and handling, tables of bearings, accessories, and spare parts; tables and drawings of parts particularly liable to wear, standards of acceptance and records, and tables of the standard elements list, and purchase list, etc., which should quote name, quantity, material etc.

19.4   TECHNOLOGICAL ANALYSIS OF THE FOOD PRODUCTION PROCESS

This analysis is the basis of food machinery design. A type of food may be produced by means of different technological processes. Taking the complexity of the physical properties of food and the varieties of food products into consideration, to analyse the technology, and to choose the optimum plan from several feasible ones, is the most important task and the key step in food machine design. In other words, the construction and cost of the designed food machine are dependent on the selected technological process.

19.4.1   Requirements of the technological process
The designed machine must meet the following requirements in the food production process.

(1)   It should ensure that the quality of food to be produced is satisfactory not only in appearance but also in taste and tactile sensation.
(2)   It should meet the preset capacity and be able to be synchronized with other equipment in the production line.
(3)   It should be convenient for the materials to be fed, shaped, and delivered.
(4)   It should be simple in construction and economical in power consumption, operation, labour, and investment.

19.4.2   Determination of the production process

19.4.2.1   *Study of the food technology before design*
It is essential for machine designers to work with food technologists to study the crux of the process, choose the optimum process and the technical parameters of the materials used at every stage of production, and present practical measures to guarantee the quality of the products before the machine is designed.

To ensure food quality it is essential to employ the latest scientific achievements and modern techniques so that conventional technology may be changed and the production process simplified. This can allow mechanization of the system and process automation.

19.4.2.2   Analysis of the food process

Food processing is the manner in which raw food materials are converted into food products by means of machinery. The wide range of food materials and product varieties make the production process different for each process. To analyse the food process is to analyse the primary food materials and the products at every stage for their properties and characteristics and use in production. After a scientific comparison the most reasonable production method and operation procedure may be chosen, and the food production process is determined.

According to their physical properties, food materials fall into four types: powder, fluid, plastic, and viscoelastic materials. Different food materials correspond to different technological processes and different technical factors.

Flour, sugar, and milk powder are typical powder materials. They have a certain flowability, inner and outer friction factors, and different densities under different handling conditions, which should all be taken into consideration in the processing machine design.

Syrup and sponge batter are liquid materials. Viscosity is the most important factor here in the fluid delivery and depositing process.

Candy massecuites, and soft and short doughs, are plastic materials which are suitable for rotary moulding, extruding, or wire-cutting process such as cookies. The candy process imparts force resulting in plastic deformation of the material with no or very little elastic deformation. Viscosity is also a key factor which is closely related to the stress and shear rate for the required deformation.

Hard doughs for biscuits and Chinese snacks and fermented doughs for bread are typical viscoelastic materials, and are perhaps the most complicated, as they show both solid and fluid behaviour (elasticity and viscosity) when subjected to a sudden instantaneous, constant shear stress. To prevent them from showing pure elasticity, sufficient time and enough stress should be provided [43], which should be determined before the machine is designed.

To shape plastic and viscoelastic materials into specific products cannot simply be done by borrowing the method used to shape metal products. For instance, in the Chinese moon cake formation process, employing the 'deep die depress drawing process' used to shape metal containers, has resulted in an undesirable design, while the Rheon encrusting process (Chapter 13) leads to a successful design since its shaping process is based on food rheology. This helps us to understand why the dough sheet should rest before the rotary or reciprocating cutters, why the hard dough piece cannot keep a complex pattern in biscuit production, and why the pressure should be kept for a while after folding the sheet in the Hun Tun-making process.

19.5   FOOD MACHINE CONTROL SYSTEMS

The mode of movement of the food machine is defined by the process of food production. But the ways for controlling a defined mode of movement can be different. There are four main types of control system: the transmission coordinated control system, the time control system, the travel limit control system, and the numerical program control system.

19.5.1   Transmission coordinated control system

This means that the movement of the executive mechanisms (working members) of the machine is controlled by a transmission system, while the regulators on the executive mechanisms are manual. This control system is relatively simple and easy to operate, but the degree of automation is not high. Machine reliability is mainly dependent on operator skill. Most bakery and snack machines employ this type of control system.

19.5.2   Time control system

Systems which are able to ensure that all working members (operating mechanisms) are coordinated according to time are referred to as time control system. The cam shaft is the most typical time control system. It coordinates motion sequences and characteristics (displacement, velocity, and acceleration) for every driven working member by means of a series of cams. Some cam shafts do not directly drive the operating mechanisms but drive, via cams, switches which give orders to control the action of the working members. In this case, the system can control only the action order, while the motion characteristics of the working members have to be accomplished by other devices. The time control system is so characterized that it is possible to provide complicated working cycles and to meet the requirements of different modes of motion by different cam curve designs. It ensures that the working cycle can be carried out steadily and precisely at the proper time. Having been regulated at the beginning there will be no interference between the working members even if each driven shaft does not run at the same speed. However, it is not very flexible. For different products, it is often necessary to change the corresponding cams. The Hun Tun maker and the Rheon encrusting machine are equipped with this type of control system, and each programme design is worked out by means of working cycle diagrams.

The working cycle diagram is the basis of cam distribution shaft design, assembling, and regulating. Fig. 19.1 shows the working cycle diagram of a typical Hun Tun maker.

19.5.3   Travel limit control system

This system ensures that the required displacement is coordinated according to the position of the working member. Generally, it works in the control mode of order–answer–order. In other words, it employs limit switches or other similar methods such as transducers to give instructions to make the working member send an answer signal when it reaches the preset displacement and this is used as the instruction for the next action. Therefore, if it fails to finish the whole programme, the operation will stop half way without mishap. Because of this, the limit control system is safer than the time control system. But since the whole operation has not yet been finished, the succeeding working members are not able to act and this would affect the food materials. This kind of system is not suitable for use in circumstances where there are high temperatures or possible pollution from oil, water, or dust, which may affect the precision action of the limit switches. Beating and mixing machines often employ this kind of control system. For instance, mixing cannot be started unless the bowl is in position, and the mixing bowl cannot be lowered until the mixing tool is stationary.

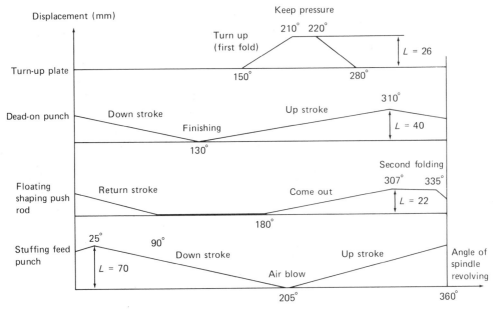

Fig. 19.1   Working cycle diagram of a typical Hun Tun maker.

19.5.4 Numerical program control system

This is a system which controls the motion of working members of the machine completely by means of the program stored in the computer. Modern control systems are also equipped with a set of integrated signal feedback systems, which can send feedback signals when any deviation of the working members takes place. The computer will immediately make suitable amendment for the deviation in response to the feedback signal so that the machine can then be restored to its normal condition. This control system is highly automatic, it costs more than other systems, and it needs greater skills for maintenance. Modern advanced electric ovens are often equipped with this kind of system to control the baking temperature, band tension, and tracking, etc.

19.6 TRANSMISSION SYSTEM DESIGN

The transmission system is used to connect the drive with the working members, to link one operating mechanism with another, to transfer the necessary power for food processing and to ensure the correct motion relationship among the working parts. The method of transmission is closely related to the machine function, construction, and manufacturing cost, so that attention should also be paid to its design in order to meet feasible and economical requirements for the machine.

   The theoretical analyses and calculations of transmission systems in food machinery design are the same as those of traditional machine tool design. It is unecessary, and indeed impossible, to cover it thoroughly in the limited space of this book. Here it should be mentioned that most food processing machines, unlike other machine tools, must be designed with a stepless vari-speed drive for synchronizing with other equipment in the food production line such as the biscuit line, bread line, candy line, etc. The application of a stepless vari-speed drive allows the machine's working parts to operate at any desired speed, which ensures high quality products, reasonable power efficiency, and variations in production capacity. Speed changing can be carried out by either a vari-speed electric motor, pneumatic device, hydraulic device, or a mechanical stepless vari-speed drive which is the most popular for food processing machines such as beating machines, rotary moulders, and rotary cutters.

Stepless vari-speed drive, or infinitely variable drive, is commonly designed with a V-belt riding on two expandable pulleys, as shown in Fig. 19.2. Each pulley is made of two tapered disks whose working surfaces form a groove for the V-belt. The variation of speed is carried out by expanding one pair of disks and contracting the other pair via a handwheel. 'About $\frac{2}{3}$ of the mechanical stepless drives manufactured in the World are of the V-belt type.' [33] The belts are often cogged (toothed) on the inner, outer, or both surfaces (Fig. 19.2). The range of speed variation is dependent on the width of the belt. Standard V-belts provide a range of speed variation from 1.11 to 3.2 (Table 19.1). Special wide belts permit a range up to 9 or even 12, (in Table 19.2) up to 6.76.

In the selection of the belt type and in the calculation for the appropriate dimensions of the stepless drive, the following should be considered.

(1)   The power and speed of the electric motor.
(2)   The type of machine.
(3)   The maximum and minimum speed of the driven pulley.
(4)   The maximum and minimum centre distance between the two pulleys.
(5)   The length of operating time.

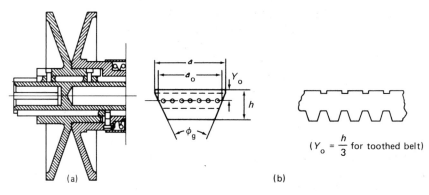

$(Y_o = \dfrac{h}{3}$ for toothed belt)

(a)   (b)

Fig. 19.2   Stepless vari-speed drives, (a) Expandable tapered-disk pulley (b) V-belt. See Table 19.1 for a_o, Y_o, h, and ϕ_g. (From D. N. Reshetov, *Machine design*, translated by Nicholas Weinstein L. (1978), English translation copyright © Mir Publishers, Moscow, (1978).

Table 19.1　Transmission parameters of the standard V-belt for stepless drive [15]

Model of belt	O	A	B	C	D	E	F
Y_o (mm)	2.1	2.8	4.1	4.8	6.9	8.3	11.
a_o (mm)	8.5	11	14	19	27	32	42
h (mm)	6	8	10.5	13.5	19	23.5	30
a_o/h	1.416	1.375	1.33	1.407	1.421	1.36	1.40
Min. diameter of adjustable pulley d(mm)	63	90	125	200	315	500	800
Groove angle of pulley ϕ_g (°)	34	34	34	34	36	36	36
d/h	10.50	11.25	11.90	14.7	16.6	21.3	26.7
Common $R_{sv} = D/d$ pulley for Type I	1.29	1.26	1.25	1.21	1.17	1.13	1.11
R_{sv} for Type II	1.66	1.59	1.56	1.46	1.37	1.28	1.23
Grooved R_{sv} pulley for Type I	1.79	1.71	1.66	1.56	1.47	1.35	1.29
R_{sv} for Type II	3.20	2.92	2.76	2.43	2.16	1.82	1.66
Motor speed 1450 rpm constant power transfer of single belt N (kW)	0.30	0.76	1.76	4.83	9.32		

Table 19.2　Transmission parameters of the special wide V-belt for stepless drive [15]

a (mm)	27	32			45	50			68	75	98
Wedge angle of belt ϕ (°)	32	32	32	30	30	30	40	32	36	40	40
a_o (mm)	25.1	30.1	29.7	29.7	42.7	47.1	46.1	46.1	64.1	70.6	93.1
a_o/h	2.51	3.01	2.48	2.88	3.28	2.94	2.88	2.44	3.56	3.92	4.66
Groove angle of pulley ϕ_g (°)	28	28	28	26	26	26	36	28	32	36	36
Min. pulley diameter d(mm)	67	71	80	90	100	125	125	150	140	150	170
d/h	6.7	7.1	6.7	6.9	7.6	7.8	7.8	7.9	7.8	8.3	8.5
R_{sv} for Type I	2.23	2.43	2.21	2.17	2.6	2.38	1.99	2.08	2.34	2.22	2.45
R_{sv} for Type II	4.97	5.90	4.88	4.71	6.76	5.66	3.96	4.33	5.48	4.93	6.00
$(K_o) = 20 - 4a_o/h$ (kg/cm²)	9.96	7.96	10.48	10.88	6.88	8.24	8.48	10.2	5.76	4.32	
$F = 0.01\,a$ (cm²)	2.51	3.01	3.56	3.86	5.55	7.54	7.38	8.82	11.54	12.71	
$P = (K_o) \cdot F$ (kg)	25.0	24.0	37.4	42.0	38.2	62.1	62.5	90.3	66.5	54.9	
h (mm) rev/min	10	10	12	13	13	16	16	19	18	18	20
Motor speed 1450 rev/min constant power transfer (Single belt) N (kW)	1.25	1.27	2.24	2.3	2.85	5.73	5.82	10.92	6.56	6.24	

The electric motor power and the range of speed variation, R_{sv}, are the bases for stepless drive design. R_{sv} is calculated from

$$R_{sv} = \frac{n_{max}}{n_{min}} \tag{19.1}$$

where n_{max} is the maximum speed of the operating mechanism (rev/min), and n_{min} is the minimum speed of the mechanism (rev/min).

The calculation of power N_o is given by

$$N_o = N_e K \tag{19.2}$$

where N_e is the rated power of the motor (kW) and K is the service factor shown in Table 19.3.

Here is an example showing how to design a V-belt stepless drive. The rated power of the electric motor is 1.1 kW, the maximum and minimum speeds of the operating member are 260 rev/min and 100 rev/min respectively, the daily working time is three shifts, 24 hours, and the service conditions are light load without repeated off and on (stop and start).

Given: $N_e = 1.1$ kW

$n_{max} = 260$ rev/min

$n_{min} = 100$ rev/min

Approach: Use Table 19.3 for K; $K = 1.2$

Solution: $N_o = 1.1 \times 1.2 = 1.32$ kW   (Equation (19.2))

$$R_{sv} = \frac{260}{100} = 1.6 \qquad \text{(Equation (19.1))}$$

Selection of V-belt: Use Table 19.1 for standard type I with grooved pulleys.

The range of speed variation closest to the calculated value is $R_{sv} = 1.66$

Table 19.3   Service factor (K) of belt

Service characteristics	Daily operating time (hours)		
	8	12	24
Light loading (stable loading under rated, without often off and on)	1.0	1.1	1.2
Moderate loading (the max. load is 125% of rated)	1.2	1.3	1.4
Heavy loading (the max. is 150% of rated)	1.3	1.4	1.5

From table 19.1: Minimum diameter of adjustable pulley $d = 125$ mm

Maximum diameter of adjustable pulley $D = 1.66 \times 125 = 207.5$ mm

Maximum power transferred by single standard belt $N = 1.76$ kW because $N_o = 1.32$ kW

therefore $N > N_o$.

In this case, if the special wide V-belt is to be selected, the closest value of R_{sv} is 2.21 (Table 19.2), and the power that can be transferred by a single belt is 2.24 kW which is much greater than the calculated value 1.32 kW, so that too much power is stored and a larger machine would result. Because of this, the application of the special wide V-belt in this case is not economical.

The length of the belt may be determined according to the *Mechanical design book* from which the centre distance between two pulleys can also be calculated.

If the machine requirements on speed changing can be met directly by the speed variation range of the stepless drive, a fixed ratio gearing may be used too meet the requirements of the operation speed. If the range of speed variation of the stepless variable drive is too narrow to meet the requirement of the machine, then stepped variable drives (gear box, multy-speed motor) in series with a stepless variable drive should be employed to further widen the range of speed variation, but it is necessary to ensure that there is no speed interruption.

References and bibliography

[1] Matz, Samuel A. (1972) *Bakery technology and engineering*, Second Edition, AVI Publishing Company, Inc., Westport, CT.

[2] Pyler, E. J. (1973) *Baking science and technology*, Volume II, Siebel Publishing Company, Chicago, IL.

[3] Wade, P. (1988) *Biscuits, cookies and crackers*, Volume 1, The principles of the craft, Elsevier Applied Science, London and New York.

[4] Almond, N. (1989) *Biscuits, cookies and crackers*, Volume 2, The biscuit making process, Elsevier Applied Science, London and New York.

[5] Brown, J. (1982) *The masters' book of bread making*, Turret Press Ltd, London.

[6] Ma Rui (1984) Technical conditions of the steamed-bun maker, Standard SBJ 005-xx, Second draft issued by The Ministry of Commerce of China, Beijing.

[7] Alikonis, Justin J. (1979) Candy technology, AVI Publishing Company, Inc. Westport, CT.

[8] Matz, Samuel A. & Matz, Theresa D. (1978) *Cookie and cracker technology*, Second Edition, AVI Publishing Company, Inc., Westport, CT.

[9] Japanese Food Machinery Industry Society (1980) *Food machinery of Japan*, Printed in Japan 3058-802380-2301.

[10] Japanese Food Machinery Industry Society and Japanese National Meat Industry Cooperation Group (1985) *Food machinery of Japan*, Japan.

[11] Manley, D. J. R. (1991) Technology of biscuits, crackers and cookies, Ellis Horwood, Chichester, Second edition.

[12] Zhong, Z. H. & Wu, H. X. (1986) Optimum design of biscuit-reciprocating-cutting mechanism, Academic theses of National Machinery Design Technique Application Interchange Conference, Volume V, Machinery Design & Transmission Institute of The Chinese Machinery Engineering Institute, Peking, China.

[13] Toledo, R. T. (1980) *Fundamentals of food process engineering*, AVI Publishing Company, Inc. Westport, CT.

[14] Ignatyev, N., Kakoilo, A., Khomyakov V., & Mikheyev Yu., (1982) *Machine tool design*, General editor Acherkan N., Translated from the Russian by

Nicholas Weinstein Vol. 4, Third printing, English translation, Mir Publishers. Moscow.

[15]   Cheng, L. M., Xiu, K. F., Yang, Y. Y., Wang, Z. G. & Sun, Z. H. (1988) *Food machinery*, Chinese Food Publishing House, Peking, China.

[16]   Ling, E. F. (1979) *Food machinery and equipment*, Tian Jin Light Industry Institute, Tian Jin, China.

[17]   Lu, W. K., Li, T. J., & Zhang, Z. Q. (1983) *Heating technique of far-infrared radiation*, Shang Hai Science and Technique Publishing House, China.

[18]   Singh, R. P. & Heldman, D. R. (1984) *Introduction to food engineering, food science and technology*, A series of monographs, Academic Press, Inc. New York.

[19]   Sparrow, E. M. & Cess, R. D. (1978) *Radiation heat transfer*, Hemisphere Publishing Corporation, Washington, London.

[20]   Tian Jin Light Industry Institute & Wu Xi Light Industry Institute, (1985) *Food engineering fundamentals*, Light Industry Publishing House, China.

[21]   Lu, W. K. (1983) *Far-infrared radiation heating technique*, Shang Hai Science and Technology Publishing House, China.

[22]   Charm, S. E. (1978) *The fundamentals of food engineering*, Third Edition, AVI Publishing Company, Inc., Westport, CT.

[23]   Li, G. C. (1989) *Food machinery fundamentals and applied technique*, Hei Longjiang Science and Technology Publishing House, Harbin, China.

[24]   Cheng, L. M., Xiu, K. F., Yang, Y. Y., Wang, Z. G., Liu, Z. Y., Zou, B., Shi, Y., & Wang, M. N. (1987) *Food factory planning with special reference to bakery and confectionery food*, Agricultural Publishing House, Peking, China.

[25]   Wu, M., Wang, C. Y, & Sun, J. Y. (1989) *Chinese cakes*, Chinese Commerce Publishing House, Peking, China.

[26]   Boltman, B. (1978) *Cook-freeze catering systems*, Applied Science Publishers Ltd., London.

[27]   Loncin, M. & Merson, R. L. (1979) *Food engineering, principles and selected applications*, Academic Press, New York, London, San Francisco.

[28]   Mckenna, B. M. (1984) *Engineering and food*, Volume 1, Engineering science in the food industry, Elsevier Applied Science Publishers, London and New York.

[29]   Johnson, R. C. (1979) *Optimum design of mechanical elements*, Second edition, John Wiley & Sons, Inc. New York.

[30]   Mckenna, B. M. (1984) *Engineering and food*, Volume 2, Processing applications, Elsevier Applied Science Publishers, London and New York.

[31]   Gillies, Martha T. (1979) *Candies and other confections*, Noyes Data Corporation, Park Ridge, NJ.

[32]   Gu, Q. P. (1982, I) An introduction to WHT 80-II Hun Tun Maker, Journal of Food Science, Peking, China.

[33]   Reshetov, D. N. (1978) *Machine design*, translated from the 1975 Russian edition by Nicholas Weinstein Mir Publishers, Moscow.

[34]   Ritts, D. R. & Sissom, L. E. (1977) *Theory and problems of heat transfer*, McGraw-Hill Book Company, New York.

[35] Su, H. J. (1984) *Recent low boiling candy manufacture*, Fu Wen Book House, Tai Bei, China.

[36] Zhu, Y. (1972) *Candy processing*, Jin Wen Book House, Hong Kong.

[37] Hong, Y. P. (1991) Technical conditions of hard candy maker (First draft), Special Standards of China, The Ministry of Commerce of China, Peking.

[38] Zhang, S. W. & Yang, M. D. (1988) *Bakery food*, Hei Long Jiang Science and Technology Publishing House, Harbin, China.

[39] Peng, X. D. (1979) *Design of far-infrared baking ovens*, National Food Science and Technology information Centre, The Ministry of Commerce of China, Peking, China.

[40] Tian Jin Light Industry Institute & Wu Xi Light Industry Institute, (1984) *Food technology*, Light Industry Publishing House, Peking, China.

[41] Cheng S. D. (1984, I) A research on food feed screw, *Journal of Food Science*, Peking, China.

[42] Qi, B. A. (1982, V) Inspection on the volume of steamed bun dough, *Journal of Service Science and Technology Information*, Peking, China.

[43] Sherman, P. (1970) Industrial Rheology, Academic Press Inc., London and New York.

[44] Ling, E. F. (1979) *Food machinery and equipment*, Tian Jin Light Industry Institute, Tian Jin, China.

[45] Ma, R. (1983) First draft of mixer standard of The Ministry of Commerce of China, Peking, China.

[46] Shi, Y. B. & Fan, X. M. (1985) *Food shaping principles*, Hei Longjiang Commercial College, Harbin, China.

Index